Sustainable Civil Infrastructures

Sustainable Infrastructure impacts our well-being and day-to-day lives. The infrastructures we are building today will shape our lives tomorrow. The complex and diverse nature of the impacts due to weather extremes on transportation and civil infrastructures can be seen in our roadways, bridges, and buildings. Extreme summer temperatures, droughts, flash floods, and rising numbers of freeze-thaw cycles pose challenges for civil infrastructure and can endanger public safety. We constantly hear how civil infrastructures need constant attention, preservation, and upgrading. Such improvements and developments would obviously benefit from our desired book series that provide sustainable engineering materials and designs. The economic impact is huge and much research has been conducted worldwide. The future holds many opportunities, not only for researchers in a given country, but also for the worldwide field engineers who apply and implement these technologies. We believe that no approach can succeed if it does not unite the efforts of various engineering disciplines from all over the world under one umbrella to offer a beacon of modern solutions to the global infrastructure. Experts from the various engineering disciplines around the globe will participate in this series, including: Geotechnical, Geological, Geoscience, Petroleum, Structural, Transportation, Bridge, Infrastructure, Energy, Architectural, Chemical and Materials, and other related Engineering disciplines.

More information about this series at http://www.springer.com/series/15140

Hugo Rodrigues · George Morcous ·
Mohamed Shehata
Editors

Recent Research in Sustainable Structures

Proceedings of the 3rd GeoMEast International Congress and Exhibition, Egypt 2019 on Sustainable Civil Infrastructures – The Official International Congress of the Soil-Structure Interaction Group in Egypt (SSIGE)

Springer

Editors
Hugo Rodrigues
Instituto Politécnico de Leiria
Leiria, Portugal

George Morcous
University of Nebraska-Lincoln
Nebraska City, NE, USA

Mohamed Shehata
Expertise House
for Engineering Consultant
Cairo, Egypt

ISSN 2366-3405 ISSN 2366-3413 (electronic)
Sustainable Civil Infrastructures
ISBN 978-3-030-34215-9 ISBN 978-3-030-34216-6 (eBook)
https://doi.org/10.1007/978-3-030-34216-6

This Springer imprint is published by the registered company Springer Nature Switzerland AG
The registered company address is: Gewerbestrasse 11, 6330 Cham, Switzerland

Contents

About the Editors

Hugo Rodrigues is Senior Lecturer at the School of Technology and Management, Polytechnic Institute of Leiria, Portugal, teaching several topics related to structural analysis, building pathology and rehabilitation. He received his Ph.D. in civil engineering from the University of Aveiro in 2012. His experience in seismic analysis, having participated as a team member in research and development projects, specialized consultancy studies ordered by several public institutions and companies regarding the assessment of seismic risk. His major research interests are building rehabilitation, structural health monitoring and seismic safety, including experimental and numerical activities. He is a research member of the RIsks and Sustainability in COnstruction (RISCO). He has co-authored more than 150 publications in top-tier peer-reviewed journals and national and international conferences.

Dr. George Morcous is Professor at Durham School of Architectural Engineering and Construction at the University of Nebraska-Lincoln since January 2005. He has B.S. and M.S. in civil engineering from Cairo University, Egypt. He earned his doctorate from Concordia University, Canada, in 2000. He is currently a registered professional engineer in the state of Nebraska. His research and teaching interests include design and construction of precast, prestressed concrete structures and bridge engineering. He has two patents and over 150 publications.

Dr. Mohamed Shehata; PhD, MBA, MSC, PMP, CLAQ, ASQ, CEO and founder of the EHE-Consulting Group in Middle East.

He has more than 25 years of experience in many mega, large and small projects in the Middle East. He was the leader of the multidisciplinary engineering works, so he has gained experience in the architectural master planning, urban planning, project management, project preparations, decision making and value engineering of the projects. In addition to all previous manager works, he has a professional expertise in the geotechnical, structural and bridge engineering.

Exploration of Early Civilization Through Restoring Structural Stability of DJOZER Step Pyramid (Part II)

Radwa Mahmoud[1] and Hassan F. Imam[2(✉)]

[1] Engineering at Housing Building and Research Center, Cairo, Egypt
radwamahmoud.uti@outlook.com
[2] Rock Mechanics & Underground Structures, Faculty of Engineering, Cairo University, Cairo, Egypt
hassanimam34@yahoo.com

Abstract. Step pyramid is the major archaeological unit of Djoser complex at Saqqara archaeological site. It is suffering now from various and serious deterioration and degradation features from various sources Rescue and Restoration studies of the first stone building ever witnessed i.e. pyramid of Djozer, Sakkara site, Egypt, included several aspects. Aspects under consideration in present study are concerned with geotechnical and geo environmental impacts on Djozer pyramid deterioration features of sub-surface units.

Main aspects dealt with are analysis of causes and sources of structural deterioration features of subsurface units of the pyramid itself such as galleries and entrances, burial shaft and its ceiling failure zone, and burial chamber and granitic sarcophagus. Field study plan has been carried out to diagnose deterioration features, analysis of building material, find out properties and behavior of hosting rock media, and types of structural instabilities.

Analysis of results revealed that modes of structural instability of subsurface units took form as:

- rock formation partial failure and collapses
- change of galleries section, shape and dimensions
- separation of roof rock formation
- buckling in galleries walls
- failure of rocks at the zones of intersect of galleries and vertical shaft
- Failure and deterioration of exposed swelling clay formation hosting burial shaft walls, galleries, walls of burial chamber ....etc.

1 Introduction

Saqqara archaeological site is located SW of Cairo city extending along desert about 8 km. in N-S direction with width of about 4 km.

Saqqara is considered to be one of the important archaeological areas in Egypt due to its relation to Memphis cemetery. It is really an open book that tells the whole story of the early ancient Egyptian Civilization, where it shows some architectural and engineering development never seen anywhere other than in old Egypt, presented by funerary DJOZER complex i.e. step pyramid, southern tomb and other units, which had been built using

H. Rodrigues et al. (Eds.): GeoMEast 2019, SUCI, pp. 1–15, 2020.
https://doi.org/10.1007/978-3-030-34216-6_1

stone blocks for the first time in building history. This fact represents unbelievable development in engineering design and architectural concept of royal tombs.

So dealing with both various aspects of deterioration and structural engineering analysis for restoring stability of that complex have to be tackled with serious care. All archaeological measure has to be taken into consideration to achieve structural stability of both surface and subsurface units of pyramid.

1.1 Engineering Aspects of Sub Surface Units of Pyramid

Present study is concerned of sub-surface units of step pyramid, and before going into discussing their structural deterioration, it is necessary to analysis and explains the recent documented engineering data of these units.

Study has been focused to achieve the documentation and careful identification of parameters that help in preparing engineering data base (qualitative and quantitative) for future conservation and present restoration events. It includes and discusses the influential phenomena creating structural instabilities, and rock behavior, mode of failure of rock formation that may contribute in identifying and quantifying causes of such instabilities that may be due to interaction of multi sources such as:

- Behavior of the hosting rock formation, and their swelling potentiality which is environmentally variable and hence the produced swelling pressure.
- Seasonal change of the climatic parameters, mainly humidity and temp.
- disintegration, degradation and removal of mortars (various causes)
- change of in situ stress regime, hence impact on deformation mode around burial shaft and galleries sections
- Seepage water that occurred through geologic structures in Sakkara plateau from the surrounding cultivated and reclaimed land.
- Seismic activities and human seismic- like activities.
- Weathering actions of geo- environmental and environmental parameters.
- Aging effect in general.
- Figure 1 shows 3-D of pyramid sub surface units, the relationship of various components of subsurface units of step pyramid and their relative levels.

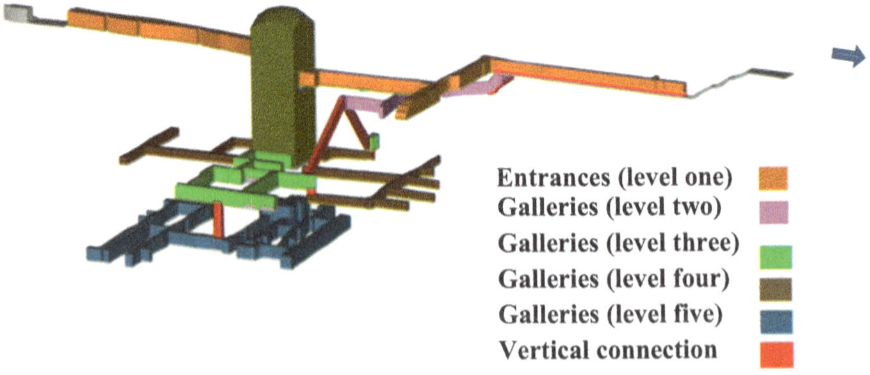

Fig. 1. 3D model of pyramid subsurface units

1.1.1 These Units Could Be Classified Archaeological Wise as Follows

- *Entrances (level one)*

Step pyramid has three known entrances, all were excavated in rock formation of Saqqara plateau. Their architectural and engineering data are, as shown in integrated plan Figs. 1, 2, 3 and 4, Table 1 summarizes main engineering data of subsurface units.

- *Burial vertical shaft. shown in* Fig. 5
- *Burial room. shown in* Fig. 6, 7, 8
- *Galleries shown in* Fig. 1

Table 1. Shows the detailed engineering and architectural data

Unit	Length (m)	Cross section shape - height	Status
North. Entrance	99.3	D-shape & Rectangle, 2.5 × 4 m	Deteriorated
South. Entrance	73.9	Rectangle, 3 × 4.5 m	Heavily deteriorated
East. Entrance	70+	D-shape, 2 × 1.8 m	Partial failure
Vertical. burial. shaft	28	Square, 7 × 7 m	Partial failure
Burial roo. chamber	9.2	Square, 10 × 10 m	Partial failure
Galleries	593	D-shape & Rectangle 1.8 × 2.1 m	Deteriorated
Granitic sarcophagus	5.35	Rectangle 5.35 × 3.47 × 4.73 H	Deteriorated

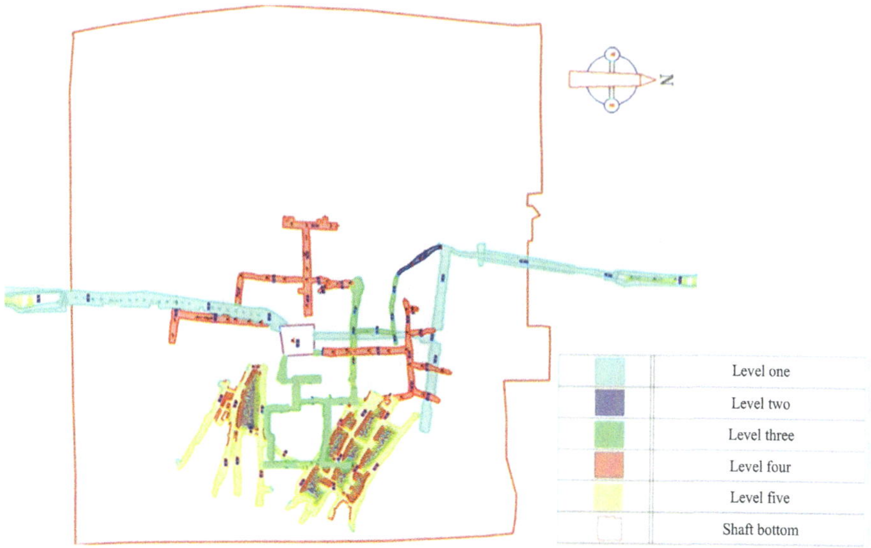

Fig. 2. Subsurface units integrated plan

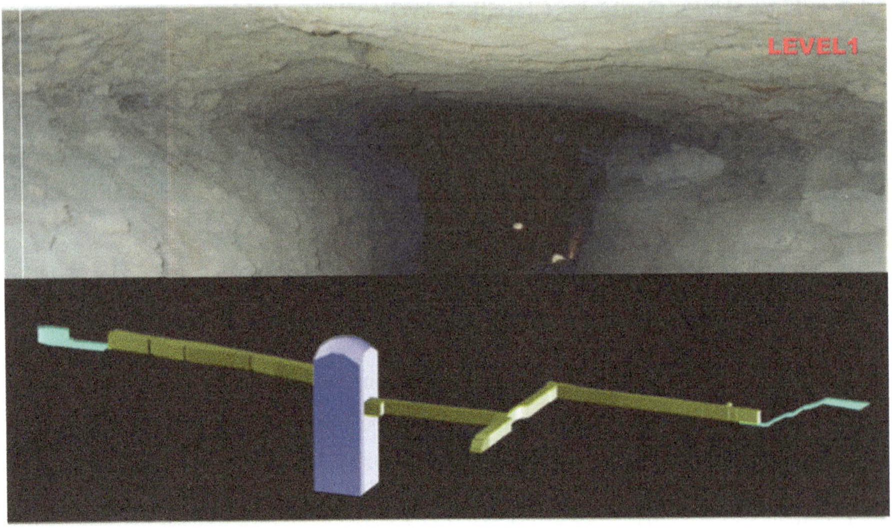

Fig. 3. Northern entrance of the Pyramid – level 1

Fig. 4. The critical state of roof layers in the southern entrance roof– level 1

Fig. 5. Burial shaft failure zone

Fig. 6. Burial room when removing debris

Fig. 7. Burial room after removing debris

Fig. 8. Granitic sarcophagus after removing debris

1.2 Analysis of Structural Instability of Subsurface Units, Causes and Sources

1.2.1 Modes of Instability

Details of field documentation survey revealed that structural instability due to various sources has the following modes of instability:

- partial failure and collapses
- deformation of galleries section, shape and dimensions
- separation of roof rock formation layers
- buckling and flacking in galleries rock wall
- failure of rock side wall at the zone of intersection of galleries and vertical shaft
- Failure of exposed layer of swelling clay formation hosting burial shaft walls, galleries, burial chamber ….etc.
- Figures 9, 10 and 11 shows some of these modes of failure and partial failure.

Fig. 9. Separation and failure of roof formation in galleries

Fig. 10. Ceramic wall panel

Fig. 11. Gallery ceramic panel, partial failure

1.2.2 Probable Causes and Sources

Field survey (FS) and results of experimental tests (ER) revealed that the most probable sources that contribute in causing such instability, degradation and deterioration as far as the results analysis of available collected data, could be as follows:

- Impact of several geo-environmental parameters, mainly humidity, temperature and subsurface seepage water (FS).
- Due to geotechnical properties and behavior of rocks where the hosting rock formation is clayey formation, parts of it have swelling potentiality (ER).
- Effect of geologic joints system on rock mass and its engineering behavior and properties.
- Seismic activities.
- Human activities.

1.2.3 Geologic Section of Saqqara Plateau

Geologic formations hosting step pyramid subsurface unites have been investigated through drilling borehole. Core description and analysis shown in Fig. 12. Rock formation that host part of units at certain level is characterized by specific colour.

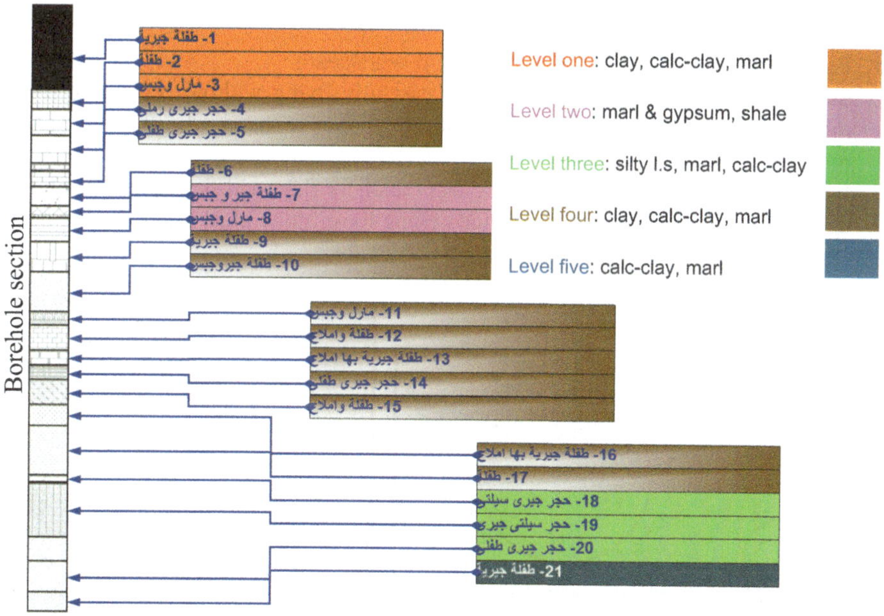

Fig. 12. Geologic section of Sakkara plateau

2 Strategy for Engineering Rescue

Due to the serious situation that the pyramid facing, and based on integrated analysis of field and testing collected data, the engineering concept of rescue and restoration may be based on the following criterions that can be carried out in parallel or in series:

- Remove all debris and insure safety of archaeological units and working team, then:
- re stabilization of burial shaft ceiling i.e. collapsed zone (1st stage)
- re stabilization of burial shaft walls (2nd stage)
- re stabilization of burial chamber and sarcophagus (3rd stage)
- re stabilization of galleries and ceramic panels (4th stage)

2.1 Concepts for Rescue and Restoration Phases

Analysis of field and laboratory investigations have shown that pyramid construction plan suffer changes several times, and that the pyramid's current form is the resultant of a long process of development that included both experimentation and improvised elements. Step pyramid is actually considered to be the evolutionary basis of all later pyramids in Egypt, rescue plan and structural restoration of heavily deteriorated zones have been proposed as follows:

- *shaft collapsed zone*

1. First insure safety of both monument and work team.
2. Stabilization of ceiling loose stones by carefully fixing with no cementing mortar mix for depth ranging between (25–50 cm) depth replacing disintegrated and/or removed mortar.
3. Investigating the suitability of using steel anchors (lab. and field experimentation) and examine its suitability to combine ceiling loose stone blocks and to what depth.

Fig. 13. Steel anchors to support shaft failure zone

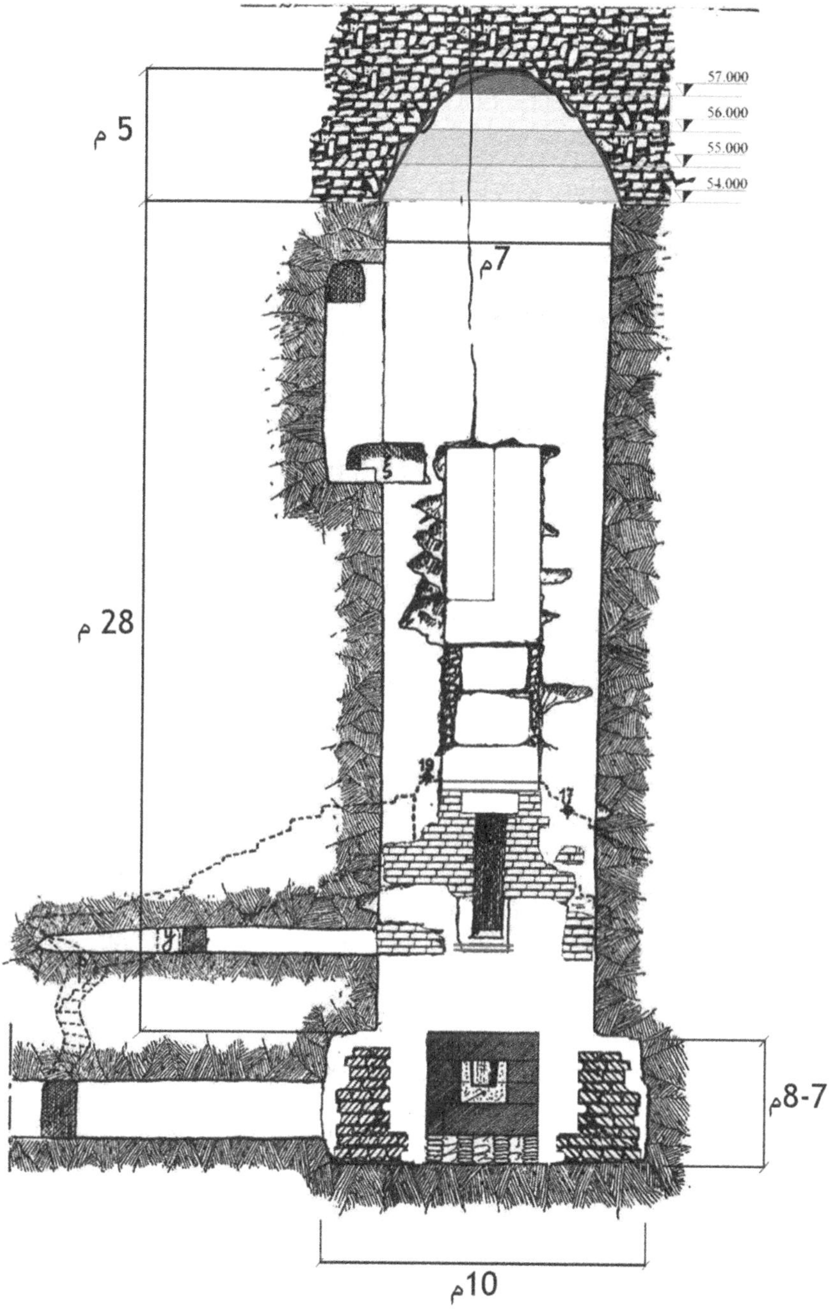

Fig. 14. E-W section of shaft and deterioration features

4. if unsuitable, injection of suitable natural materials mix will be the alternative solution to create required bonding and stability of stone blocks,
5. If steel anchors are suitable, supporting will continue, and for more safety injection for less depth will follow, Fig. 13.
6. All archaeological measures have been considered whenever possible.

- *Burial Vertical shaft walls*

Figure 14 shows northern wall of vertical burial shaft and all deterioration features including zone, Also the granitic sarcophagus shown as the bottom.

- **Burial chamber and granitic sarcophagus**

1. first insure safety of both monument and workers
2. Removal of debris produced from various partial failures and covered the chamber and all the archaeological contents.
3. Re-discover the granitic sarcophagus (weight 176 ton) and remains of archaeological limestone walls.
4. Remains of three mummies were found embedded in debris.
5. Once reached burial chamber rock floor, insuring stabilization and safety for the sarcophagus started immediately, meanwhile documentation, scientific and engineering studies of status of chamber started, results were considered in restoration project modification.
6. Radiogram survey has been carried out on rock floor of burial chamber to explore the continuity of rock media where scaffold has to be erected to reach the failure zone at top of shaft.
7. Figures 6, 7 and 8 show some of steps mentioned above.

- **Galleries**

1. Galleries as shown in Fig. 1 had been excavated at five levels, connected together by inclined or vertical connection (ladder way).
2. Some of these galleries were intersected with burial vertical shaft; others were excavated at level deeper than the burial chamber.
3. Stress and deformation distribution around various sections of galleries have been studied using laboratory testing results and recent software (Flag). Based on results, proper saving and restoration methods have been adopted.

2.2 Questionable Engineering Observations

Archaic-engineering analysis of documentation data coupled with field engineering observations and history of civilizations, created many questions and revealed that many of recent technology criterions, as the authors believe, have been previously, in different discipline, practiced in early civilization!. This fact may be supported by the engineering features that observed in almost all aspects of pyramid as engineering project such as:

- Visualize architectural philosophy and religious ethics for first time in history.
- Management and Preparation for execution such as planning for labors, equipment's, material handling into and out of project site, ventilation for sub- surface work, lightening, labors safety. Such issues are of major challenge in modern engineering
- Controlling humidity and temp. during work i.e. work environment
- Planning of galleries to meet the burial shaft at several levels
- Connecting galleries by crosscuts may be to facilitate material handling and ventilation
- At the bottom end of the vertical shaft, the burial chamber started with section larger than the shaft Section 10 m × 10 m, creating rock pillar mostly of clayey formation of 1.5 m W × 28 m H. which needs careful support.
- Sarcophagus made out of 32 granitic blocks, each of average dimensions 3.47 × 5.35 × 4.73 m, from Aswan quarries.
- Total sarcophagus weight is about 176 ton, laid down at center of burial chamber and the shaft at depth 36 m. approx.
- Excavation of galleries continues below chamber level to other level to accommodate the marble sarcophagus
- Simulation and prototype
  - In terms of modeling and prototyping that belong to recent civilization, as far as the authors believe, in early Egyptian civilization ancient engineers had practiced similar engineering concept.
  - Before executing the main sub surface units of the step pyramid, old Egyptians made prototype (model) on small scale of the planned units and executed in the same level and geological formation and features in the southern tomb south to the pyramid as shown in Fig. 15.
  - Old Egyptians studied the prototype carefully and have engineering data base for most influential parameters that should be taken into consideration when executing the main pyramid sub surface constructions.
- Set of perpendicular joint system has been found influencing the hosting rock formation. authors believe that such set of rock joints are much older than the pyramid i.e. the pyramid sub surface units, for sake of analysis, are considered to be executed in jointed rock formation.

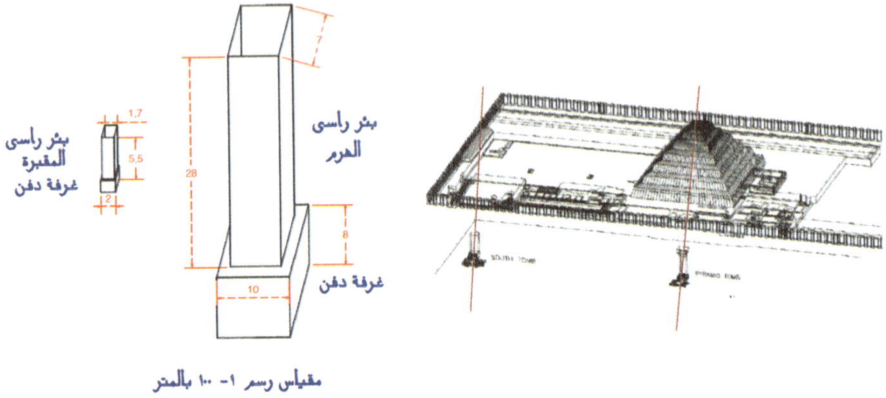

Fig. 15. Burial shafts of both pyramid and southern tomb

3 Conclusions and Recommendations

- Dealing with step pyramid deterioration from engineering view, revealed that still we need to explore deeper in early civilization aspects specially technical and cultural aspects.
- One may conclude, after analysis of engineering data of the pyramid as engineering project, that the ancient engineers were acquainted with many basic engineering fundamentals that recent civilization is practicing such as: project planning and management, modeling and prototypes, how to manage constructing in problematic soil as in swelling clay formation, and how to manage the environmental parameters seasonal changes.
- Producing of engineering and geo-environmental data base for such unique archaeological unit is crucial in order to preserve it.

موجز

مظاهر ومصادر التدهور الإنشائى للهرم المدرج

يعتبر الهرم المدرج أكبر الوحدات الأثرية بمجموعة زوسر الأثرية بمنطقة سقارة، ويعانى الهرم الآن تدهورا خطيرا يعزى الى الأنشطة الانشائية والبيئية والسيزمية ، وقبل البدء فى أى من وسائل الانقاذ والترميم كان حتميا اعداد تحليل وتقويم كامل للوضع الراهن من الزوايا الهندسية، الجيوهندسية، الجيوتقنية ، والجيوبيئية والبيئية الى جانب الأبعاد التوثيقية لنتائج هذا التقييم، وبناء على مخرجات تحليل النتائج يتم تصميم واختيار افضل البدائل لدرء الخطورة والترميم الهندسى .

Acknowledgement. Authors are indeed thankful for following archaeologists and engineers for their help and encouragement, and their fruitful archaeological and engineering discussions which made present work and its results real and possible:

* Prof. Dr. Z. Hawas

* Arch. O. Shimy, K. Waheed, & S. Farag.

* Eng. W. Abu Ela

Finally thanks are due to SCA _ PS for financing the study.

References

1. Dumas, F.: Civilization of Ancient Egypt. SCA. (1998). (in Arabic)
2. Fakhry, A.: Egyptian pyramids, the Anglo-Egyptian (2007). (in Arabic)
3. Badawy, A.: History of ancient Egyptian architecture. Part I. SCA (1988). (in Arabic)
4. Imam, H.F. et al.: Technical reports of lab. and field tests and investigations of all structural materials and rock media, including stones and mortars. SCA (2002)
5. Imam, H.F. et al.: Surveying documentation of step pyramid, both surface and inside units. SCA & Al (2002)
6. Imam, H.F., et al.: Documentation of various types and zones of deterioration, degradation and structural instabilities. SCA (2002)
7. Zang, A., Stephansson, O.: Stress Field of the Earth Crust. Springer Science Business Media (2010)
8. Hoek, E., Brown, E.: Underground Excavation in Rock. IMM (1982)
9. Revnt, I., Rode, A.: Experimental Methods of Studying Soil Structure. Amerind Pub (1969)
10. Brady, B.H.G., Brown, E.T.: Rock Mechanics For Underground Mining, 2nd edn. Chapman and Hall (1993)
11. Gramberg, J.: A Non-conventional View on Rock Mechanics and Fracture Mechanics, Rotterdam (1989)
12. Imam, H.F.: Engineering Geology and Methods of Application in Civil Engineering. Elalamya Publishing (2010). (In Arabic)
13. Lauer, J.: Saqqarq, City of the Dead (1990)

Nonlinear Finite Element Modelling for R.C Arched Beams with Openings Strengthened with CFRP

Alaa M. Morsy, Nabil H. El-Ashkar, and Aisha Jaber(✉)

Construction and Building Engineering Department,
College of Engineering and Technology, AASTMT, Abu Kir, Alexandria, Egypt
{alaa.morsy,nabilhassan}@aast.edu,
eng.aishajaber@gmail.com

Abstract. Arched beam may be defined as a curved beam having convexity upward and supported at its ends. Its major purpose is enhancing the load carrying capacity, which may come from the stiffening behavior due to membrane action. The difference in behavior between the standard straight beams and the arched beams with different convexities is investigated numerically using Finite Element (FE) software package "ANSYS". Moreover, according to the needs for making openings, the research studies the effect of opening on the ultimate load and deflection of beams, finally the research investigates the effect of strengthening the opening using Carbon Fiber Reinforced Polymers (CFRP) laminate externally or internally.

Two different experimental models (straight, and arched beams) are used for verification with the FE model. And the results show a good agreement with an accepted error. Reinforced concrete semicircular arched beam is modeled with cross section of (150 * 250) mm^2 with an inner and outer diameter 1500 and 2000 mm respectively. Full bond is assumed between the CFRP and concrete surface and between the steel reinforcement and concrete. Brick element SOLID65 and SOLID45 was used to represent concrete element and steel plate, respectively. While LINK8 and SHELL41 were used to represent steel reinforcement and CFRP sheets respectively.

A parametric study is performed to study the effect of openings in the arched beam as well as its size and shape of opening, the curvature of arch, and the CFRP strengthening effect (number of layers, size and position of the fiber) on load-deflection response, cracking and ultimate loads. The results show that arched beams have better behavior than straight beams with the same span and cross section with and without opening. Deflection has decreased and the ultimate load has increased. The shape of the openings with the same area and different aspect ratios ranges from 1 to 3 in the mid span position have no significant effect on deflection and ultimate load. However, the circular opening has the minimal deflection and the maximum ultimate load. Increasing the height of the opening or its length decreases the ultimate load and rise the deflection. The external strengthening by CFRP laminates rise the ultimate load by various percentages according to CFRP scheme used.

Keywords: Arched beams · CFRP · Openings · Finite element

H. Rodrigues et al. (Eds.): GeoMEast 2019, SUCI, pp. 16–38, 2020.
https://doi.org/10.1007/978-3-030-34216-6_2

1 Introduction

An arched beam may be defined as a curved beam having convexity upward, and supported at its ends. The shape of the arch may be circular, elliptical or parabolic and sometimes it is made up by circular arcs of several and different radii and centers. It may be subjected to vertical, horizontal or even inclined loads. In the past, arched beams had been the backbone of major buildings. The main aim of arch is to enhance the load carrying capacity, which may come from the stiffening behavior due to membrane action (Balamuralikrishnan et al. 2009).

Arch-shaped members are widely used for construction of tunnels, bridges, silos and shells. These members are not typically provided with transverse reinforcement and may thus have a brittle behavior at failure. When subjected to bending or shear, traditional design methods used for straight members are not applicable due to deviation forces developed at the curved chords carrying compression and tension, which is not always accounted in design codes (Ali et al. 2015).

In the construction of modern buildings, the use of openings in reinforced concrete beams is a common practice in non-seismic regions. These openings are used for passage of service lines such as a network of pipes and ducts which are necessary for accommodating essential services like water supply, sewage, air-conditioning, electricity, telephone, and computer network. Usually, these pipes and ducts are placed underneath the beam soffit and, for aesthetic reasons, are covered by a suspended ceiling, thus creating a dead space. Passing these ducts through transverse openings in the floor beams leads to a reduction in the dead space and result in a more compact design (Balamuralikrishnan et al. 2009).

(Bashar et al. 2017) investigated the curvature forces on arched beams the beams were tested under the action of two point loading at the top face of mid spans with two supports at the bottom face of the beams unstrengthened and strengthened (externally by CFRP laminates or internally by steel reinforcement). The results show that the use of CFRP laminates as external confinement instead of internal stirrups at the middle sector to resist curvature forces induced between reinforcing bars and concrete cover, gave approximately the same general response, but the ductility ratio was lesser.

(Ali et al. 2014) investigated the behavior and performance of reinforced concrete horizontally semi-circular curved beams with and without openings, unstrengthened and strengthened (externally by CFRP laminates or internally by steel reinforcement). The beams were tested under the action of two point loads at top face of mid-spans with three supports at bottom face.

The results showed that the presence of opening has a great effect on the behavior and ultimate load capacity of semi-circular curved beams, while the strengthening of these opening by internal steel reinforcement or external CFRP laminates will increase the ultimate load capacity and affect post-cracking behavior and mode of failure of these beams.

(Ali et al. 2010) studied the analysis of reinforced concrete horizontally curved deep beams, loaded transversely to its plane, using a three-dimensional nonlinear finite element model in the pre and post cracking levels and up to the ultimate load. It was found that decreasing the central subtended angle causes an increase in the ultimate load resisted by curved beams. Also, it was found that the effect of the internal torsion and the amount of transverse reinforcement on the ultimate load resisted by curved beams was decrease as (a/d) ratio decrease, while the effect of the flexural moment and using additional longitudinal bars as a horizontal shear reinforcement was increase when (a/d) ratio decrease.

(Al-Mutairee 2013) studied the effects of non-uniform distribution of longitudinal reinforcements on the behavior of reinforced concrete (RC) horizontally curved beams with fixed-ends under static loads to product an optimal strength of these beams without increasing the volume of longitudinal reinforcement. The results show that the effect of non-uniform distributions of longitudinal reinforcement of RC horizontally curved beams with fixed-ends is effective and can be used to improve the strength of this type of beams and its importance increases with increasing the angle of horizontal curvature (θ).

2 Research Scope and Objectives

The main objectives of the research are studying the behavior of arched beams with opening compared to straight beams as well as studying the effect of openings shape and size using finite element analysis by ANSYS software. Also to present the performance of reinforced concrete arched beams externally strengthened with CFRP using different scheme.

2.1 Methodology

This research is divided as follows:

- Beams are modelled using finite element analysis by ANSYS 12 software. Results are verified with the experimental models carried by (Balamuralikrishnan et al. 2009; Ali et al. 2015).
- A parametric study is performed to study the effect of the shape and size of openings, the effect of CFRP wrapping number of layers, and CFRP externally bonded development length, as shown in Fig. 1.

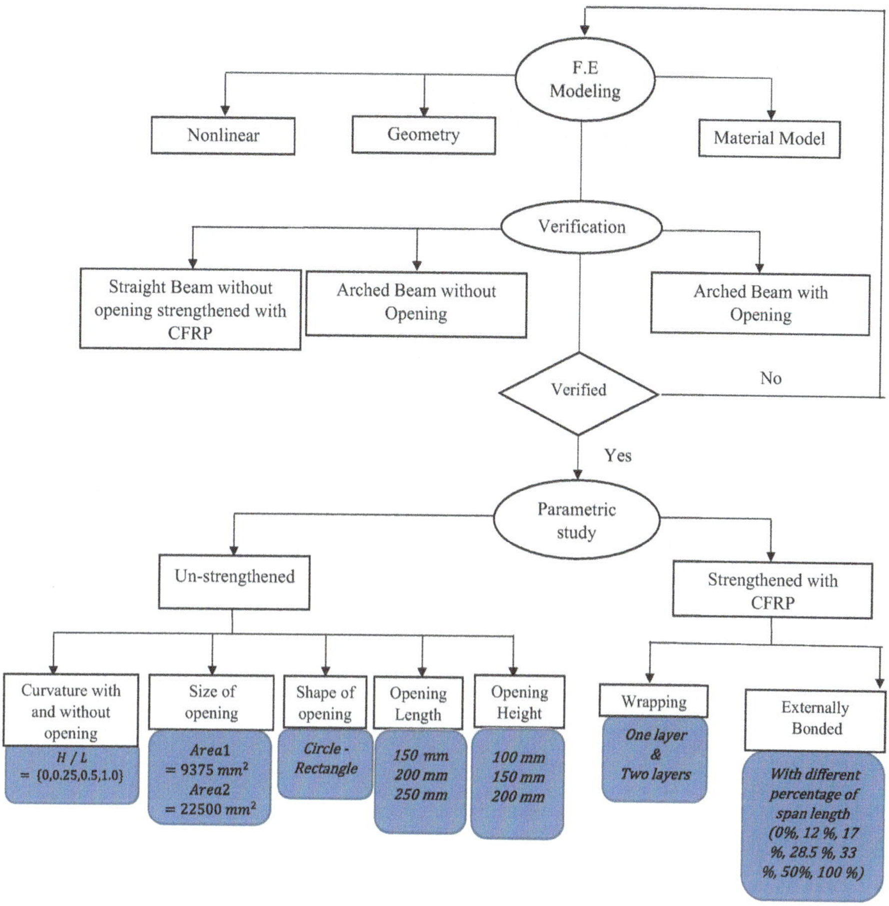

Fig. 1. Research methodology

3 Finite Element Model

3.1 Material Model

3.1.1 Modeling of Concrete

SOLID65 is used for the 3-D modeling of concrete. The element is defined by eight nodes having three degrees of freedom at each node i.e. translations in the nodal x, y and z directions. The most important aspect of this element is the treatment of nonlinear material properties (Guide 2011). The concrete is capable of cracking (in three orthogonal directions), crushing, plastic deformation and creep. Concrete was assumed to be both linear elastic and multi-linear inelastic material. Figure 2 Stress Strain Curve

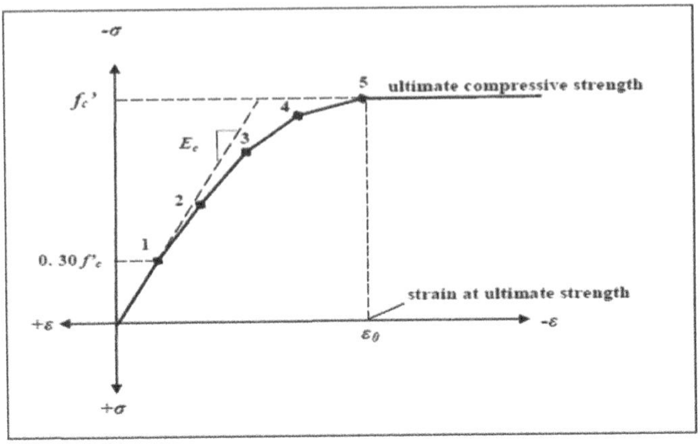

Fig. 2. Stress strain curve of concrete

for Concrete. Compressive strength of concrete was 33 MPa Poisson's ratio of 0.2 was used. The constant of the uniaxial stress strain curve of concrete in this study

$$f = \frac{E_c \epsilon}{1 + \left(\frac{\epsilon}{\epsilon_0}\right)^2}$$

$$\epsilon_0 = \frac{2f'_c}{E_c}$$

$$E_c = \frac{f}{\epsilon}$$

Where: $f =$ stress at any strain ϵ, psi
$\epsilon =$ starin at stress f
$\epsilon_0 =$ strain at the ultimate compressive strength f'_c

3.1.2 Steel Plate

SOLID45 is used for the 3-D modeling of steel plate. The element is defined by eight nodes having three degrees of freedom at each node i.e. translations in the nodal x, y and z directions. The element has plasticity, creep, swelling, stress stiffening, large deflection and large strain capabilities (Guide 2011). Steel plate was assumed to be linear elastic material. An elastic modulus equal to 200 GPa and Poisson's ratio of 0.3 was used.

3.1.3 Steel Reinforcement

LINK8 is defined by two nodes which has used for the modeling of reinforcing bar. The 3-D spar element is a uniaxial tension-compression element with three degrees of freedom at each node: translations in the nodal x, y and z directions. As in a pin-jointed structure, no bending of the element is considered. Plasticity, creep, swelling, stress stiffening and large deflection capabilities are included (Guide 2011). Reinforcing bars are assumed to be both linear elastic and bilinear inelastic material. Figure 3 shows stress strain curve of steel. The deformed bars have (520, 470 and 525) MPa yield stress for bar diameters (6, 10 and 12) mm respectively. Poisson's ratio of 0.3 is used.

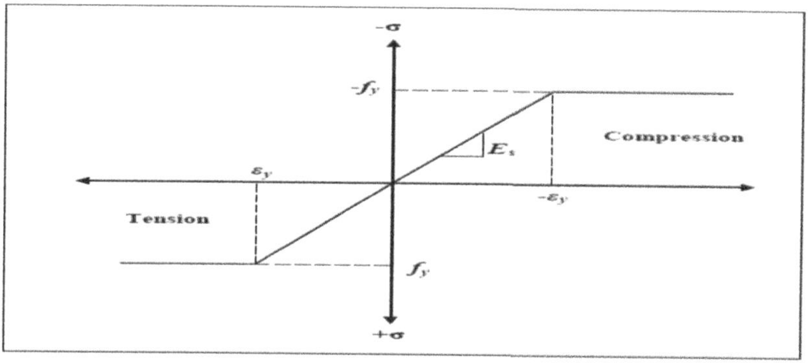

Fig. 3. Stress-strain curve of steel

3.1.4 Modelling of CFRP

SHELL41 is a 3-D element having membrane (in-plane) stiffness but no bending (out-of-plane) stiffness. The element has three degrees of freedom at each node: translations in the nodal x, y, and z directions. The geometry, node locations, and the coordinate system for this element are shown in Fig. 4. The element is defined by four nodes, four thicknesses, a material direction angle and the orthotropic material properties. Orthotropic material directions correspond to the element coordinate directions The thickness is assumed to vary smoothly over the area of the element, with the thickness input at the four nodes. A CFRP sheet has a tensile strength of 4300 MPa, and modulus of

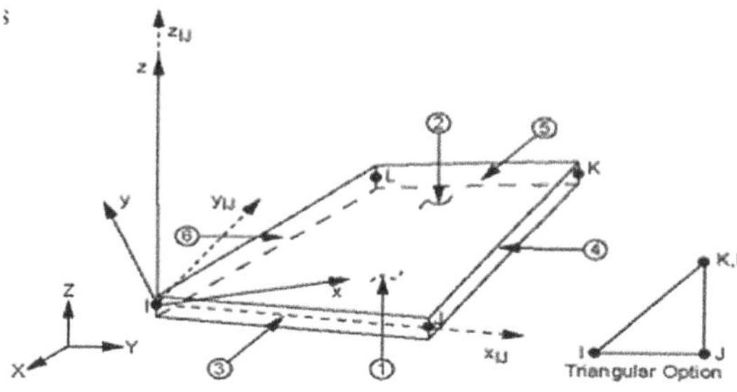

x_{IJ} = Element x-axis if ESYS is not supplied.

x = Element x-axis if ESYS is supplied.

Fig. 4. Element geometry SHELL41

elasticity of 238000 MPa, the elongation at break of 1.8% and the thickness of 0.131 mm.

3.1.5 Loading and Boundary Condition

To ensure that the model behave the same way as the experimental beam boundary conditions were needed to be applied at nodes in the supports. The supports were

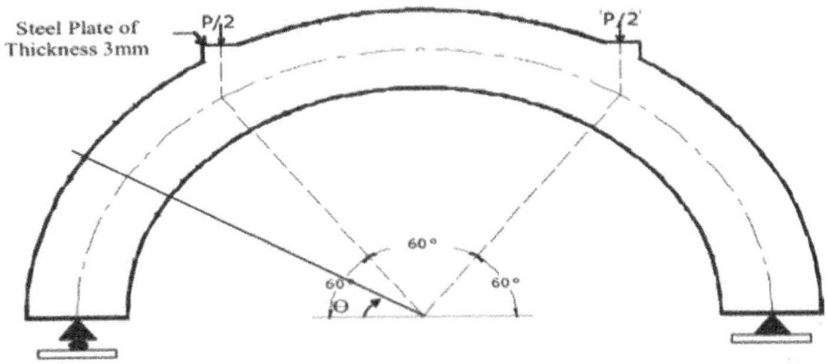

Fig. 5. Loading system on the beam

modeled to create roller and hinged supports. The force P = 180 kN was applied on all nodes through the entire centerline of the steel plates Fig. 5 shows details of loading on arched beam.

3.1.6 Nonlinear Solution and Meshing

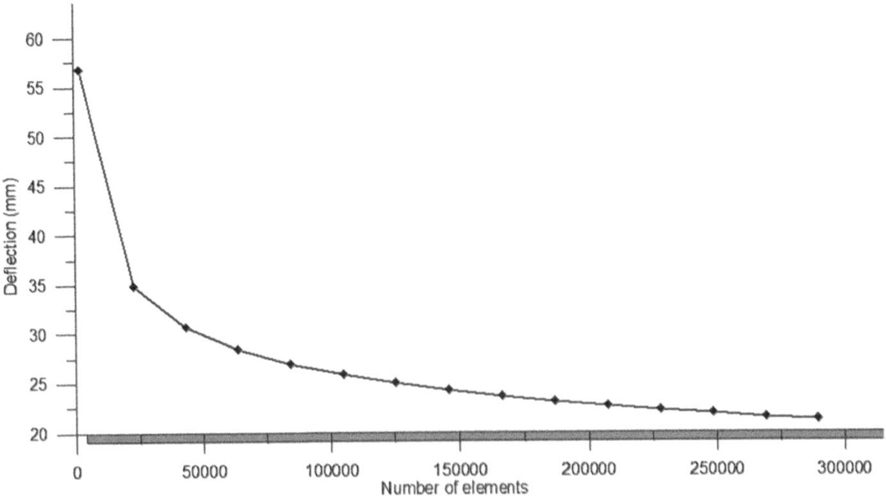

Fig. 6. Mesh independent study

Mesh independence study is performed where various number of elements ranging from about 2000 to 300000 are used. The deviation between the results for the mid span deflection obtained for 60000 elements and 80000 were less than 7%, as shown in Fig. 6. Accordingly, the 60000 elements are found to be a good compromise between the computational time and the accuracy of the solution.

4 Model Verification

4.1 First Model Verification

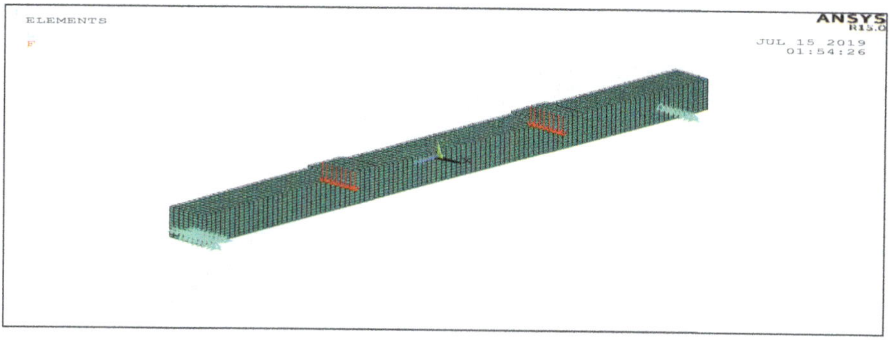

Fig. 7. Element discretization, loading pattern and boundary conditions for straight beam

The experimental investigation of (Balamuralikrishnan et al. 2009) have been studied over a controlled beam. The beams are modelled as volumes. The model is 3200 mm long, with a cross-section of 125 mm × 250 mm without FRP is called as control beam 1. The reinforcement at the bottom of beam is 2–12 mm diameter and the

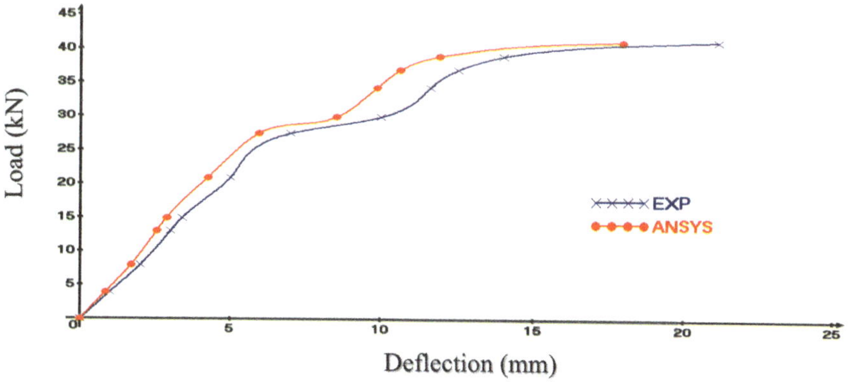

Fig. 8. Model verification of the deflection of the straight beam

reinforcement at top of beam is 2 reinforcing bars with 10 mm diameter, 6 mm diameter stirrups @ 150 mm c/c. Figure 7 shows element discretization, loading pattern and boundary conditions for straight beam.

The general behavior of the finite element model shows good agreement with observations and data from the experimental beam test. Figure 8 shows relation between ultimate load and deflection for Exp. & ANSYS model, the percentage error of ultimate load doesn't exceed 11%; which ensures result validation.

4.2 Second Model Verification

The experimental models include two simply supported RC semicircular arched beams with and without opening. (Ali et al. 2015). The arch has inner diameter of 1500 mm

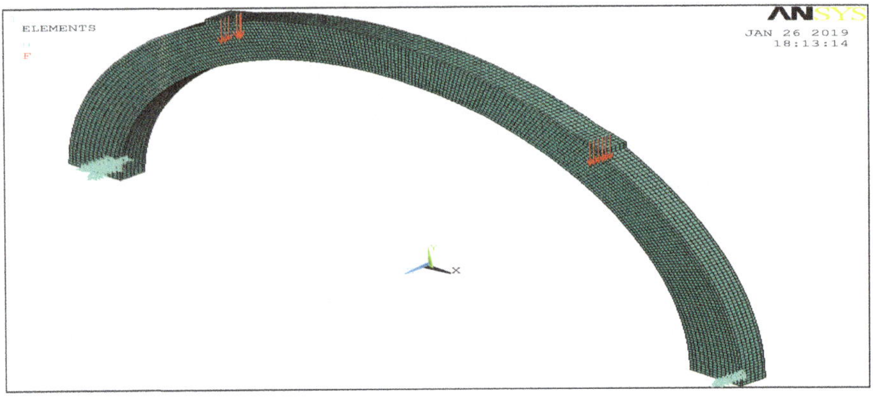

Fig. 9. Element discretization, loading pattern and boundary conditions for arched beam without opening

and outer diameter 2000 mm, and has cross section dimensions of 250 mm overall depth and 150 mm width. The arch is tested under two point loads at extrados (top) surface. The geometrical details of arches and the steel reinforcement provided with a clear concrete cover to the reinforcement of 25 mm. 2 Ø12 mm deformed bars are provided as tension reinforcement and 2 Ø10 mm top deformed bars to fasten the stirrups. The arch is provided with closed stirrups of Ø6 mm spaced at 6.25° along the circular path (95 mm along arch center line). Figure 9 shows the element discretization, loading pattern and boundary conditions for the arched beam.

In order to compare Exp. model with ANSYS, numerical results are extracted from ANSYS models, and values obtained from both approaches are given in Fig. 10. Good agreement between values are obtained from both approaches; which ensures result validation with an error percentage of ultimate load doesn't exceed 10%.

The third model is similar to the second one except an opening in its mid span with dimensions of 200 * 100 mm as shown in Fig. 11.

The general behavior of the finite element model shows good agreement with observations and data from the experimental beam test. Figure 12 shows relation

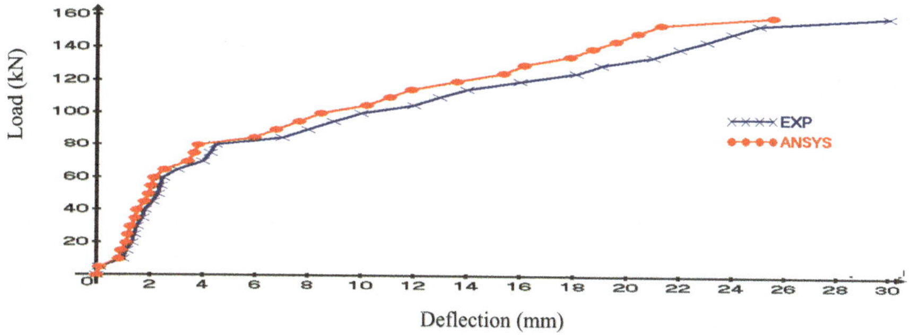

Fig. 10. Model verification of the deflection of the arched beam without opening

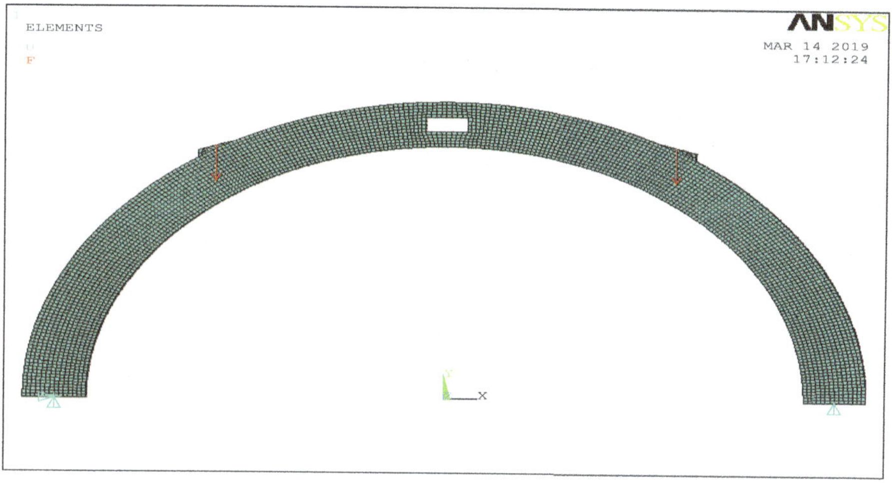

Fig. 11. Element discretization, loading pattern and boundary conditions for arched beam with opening

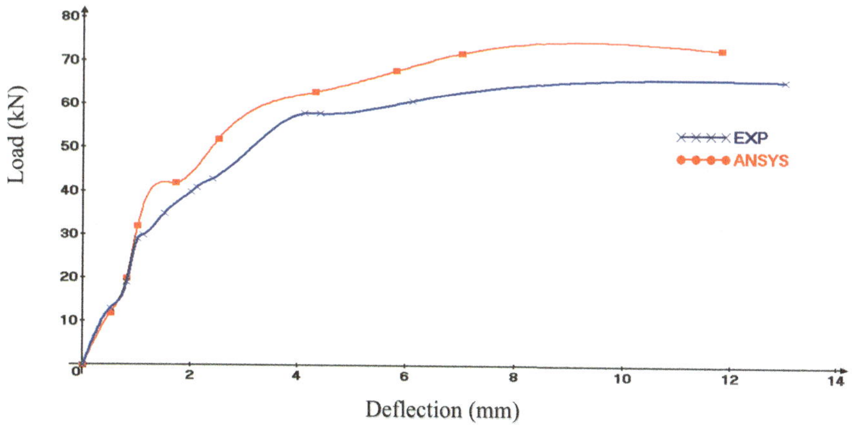

Fig. 12. Model verification of the deflection of arched beam with opening

between ultimate load and deflection for Exp & ANSYS, the percentage error of ultimate load doesn't exceed 11%; which ensures result validation.

5 Parametric Study

The ANSYS model includes simply supported RC semicircular arched beam without opening. The beam is of inner diameter 1500 mm and outer diameter 2000 mm, and has cross section dimensions 250 mm overall depth and 150 mm width. The beam is tested under two point loads at extrados (top) surface.

Table 1 indicates all the beams presented in this study (symbols, dimensions and conditions). Figure 13 shows the geometrical details of arch and the steel reinforcement provided with a clear concrete cover to the reinforcement of 25 mm. 2 Ø12 mm

Table 1. Details of all different models

Parametric study	Symbol	Height (mm)	Approximate opening dimensions				Number of CFRP layers	Percentage of CFRP from beam length
			Shape	Dim. (mm) $h_o * L_o$	Aspect ratio	Total opening area mm^2		
Curvature	B0	0	NA			NA	NA	NA
	B1	437.5	NA			NA	NA	NA
	B2	875	NA			NA	NA	NA
	B3	1750	NA			NA	NA	NA
	*B0**	0	Rectangle	86.6 * 259.8	3	22500	NA	NA
	*B1**	437.5	Rectangle	86.6 * 259.8	3	22500	NA	NA
	*B2**	875	Rectangle	86.6 * 259.8	3	22500	NA	NA
	*B3**	1750	Rectangle	86.6 * 259.8	3	22500	NA	NA
Opening size & shape	B4	875	Square	96.8 * 96.8	1	9375	NA	NA
	B5	875	Circle	D = 110	1	9375	NA	NA
	B6	875	Rectangle	79.1 * 118.6	1.5	9375	NA	NA
	B7	875	Rectangle	68.5 * 136.9	2	9375	NA	NA
	B8	875	Rectangle	61.2 * 153.1	2.5	9375	NA	NA
	B9	875	Rectangle	55.9 * 167.7	3	9375	NA	NA
	B10	875	Square	150 * 150	1	22500	NA	NA
	B11	875	Circle	D = 169.2	1	22500	NA	NA
	B12	875	Rectangle	122.5 * 183.7	1.5	22500	NA	NA
	B13	875	Rectangle	106.1 * 212.1	2	22500	NA	NA
	B14	875	Rectangle	94.9 * 237.1	2.5	22500	NA	NA
Opening length	B15	875	Rectangle	150 * 200	1.3334	30000	NA	NA
	B16	875	Rectangle	150 * 250	1.6667	37500	NA	NA
Opening height	B15*	875	Rectangle	100 * 150	1.5	15000	NA	NA
	B16*	875	Rectangle	200 * 150	1.333	30000	NA	NA
CFRP wrapping	B17	875	Rectangle	106.1 * 212	2	22500	1	NA
	B18	875	Rectangle	106.1 * 212	2	22500	2	NA
CFRP externally bonded at beam soffit	B19	875	Rectangle	106.1 * 212	2	22500	1	12.17%
	B20	875	Rectangle	106.07 * 212	2	22500	1	17.14%
	B21	875	Rectangle	106.07 * 212	2	22500	1	28.6%
	B22	875	Rectangle	106.07 * 212	2	22500	1	33.1%
	B23	875	Rectangle	106.07 * 212	2	22500	1	100%

• h_o opening height, L_o opening width

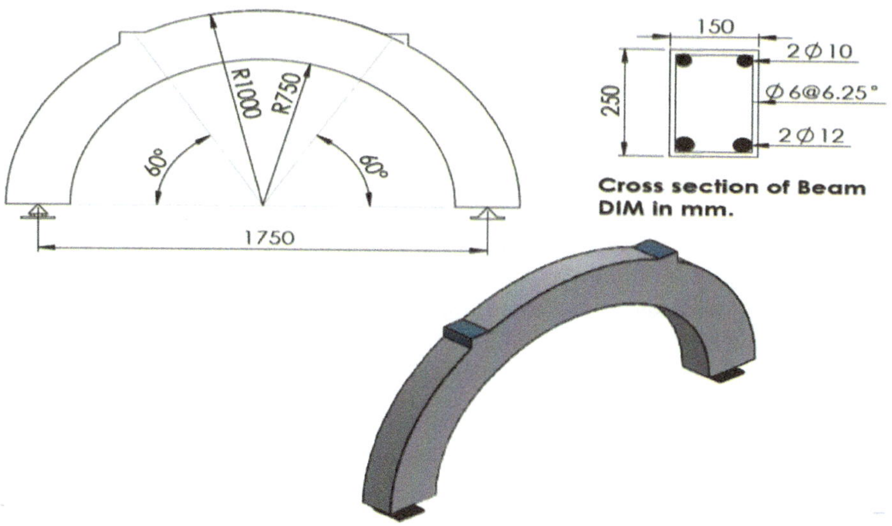

Fig. 13. Finite element model

deformed bars are provided as tension reinforcement and 2 Ø10 mm top deformed bars to fasten the stirrups. All arches are provided with closed stirrups of Ø6 mm spaced at 6.25° along the circular path (95 mm along arch center line), as shown in Fig. 13, f_{cu} = 33 MPa.

5.1 Effect of Curvature

Eight beams are modeled to study the effect of curvature on ultimate load capacity and deflection of arched concrete beams with and without openings. Four different height to

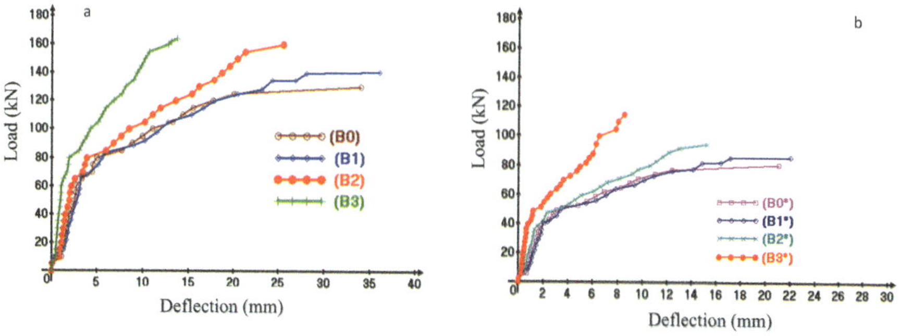

Fig. 14. Effect of curvature on arched beams with and without openings a. Without opening b. With opening

span ratio are studied; H/L = (0, 0.25, 0.5 and 1), where H is the height of the beam and L is the span. The opening size has height of 86.6 mm and length of 259.8 mm. All beams are modeled with the same cross section (250 * 150) mm^2 and span of 1750 mm under two point loads at extrados (top) surface with ultimate applied load of 90 kN at each point. The load-midspan deflection curves for all types are shown in Fig. 14.

A significant variation in general response and ultimate load are noticed. An increase in ultimate load of (8.5%, 23%, 26%) in curved beams without opening of H/L (0.25, 0.5 and 1.00) respectively compared to the straight beam (B0). and An increase in ultimate load of (6.7%, 39.7%, and 42.7%) in curved beams with opening of H/L (0.25, 0.5 and 1.00) respectively, compared to the straight beam with opening (B0*), as shown in Table 2. The percentage of increase in ultimate load is more significantly in beams with opening than without opening. This large increase in percentage is due to

Table 2. The curvature effect of beams with and without opening on ultimate load

Beam symbol	Dimensions (mm)		Ultimate load (KN)	Increase of ultimate load % compared to straight beam
	Shape of opening	Opening size $h_o * L_o$		
B0	NA	NA	130	–
B1	NA	NA	141	8.5%
B2	NA	NA	160	23%
B3	NA	NA	164	26%
*B0**	Rectangle	86.6 * 259.8	80.6	–
*B1**	Rectangle	86.6 * 259.8	86	6.7%
*B2**	Rectangle	86.6 * 259.8	112.658	39.7%
*B3**	Rectangle	86.6 * 259.8	115	42.7%

the large drop of ultimate load in the straight beams from 130 kN in the without opening case to 80.6 kN in the opening case of rectangle with aspect ratio of 3. There is a small variation in the ultimate load between the circular beams and the parabolic beam of H/L = 1. However, the height of this parabolic beam is much higher than the circular one. Then it is concluded that the circular arched beams are suitable for the rest of the study.

5.2 Effect of Opening Size

The effect of the opening size on the load-deflection curve and ultimate load capacity in a circular arched beam with two different areas of opening is studied. The smaller area equals 9375 mm^2 is presented in two different shapes: circular opening of diameter 110 mm (B5) and a rectangular opening of dimensions (55.90 mm * 167.71 mm) (B9). The larger area equals 22500 mm^2 is also presented in same two shapes: circular opening of diameter 170 mm (B11) and a rectangular opening of dimensions

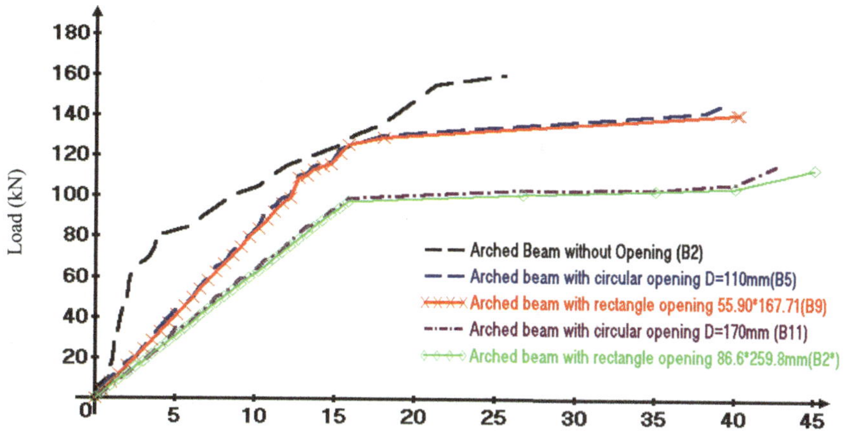

Fig. 15. Effect of opening size

Table 3. The effect of opening on ultimate load

Beam symbol	Dimensions (mm)		Ultimate load (KN)	Decrease of ultimate load % compared to B2
	Shape of opening	Opening size $h_o * L_o$		
B2	Non	Non	160	–
B5	Circle	D = 110	145	9.4%
B11	Circle	D = 170	116	27.5%
B9	Rectangle	55.90 * 167.71	141	11.87%
B2*	Rectangle	86.60 * 259.81	112.658	29.6%

(86.60 * 259.81) (B2*). Figure (15) shows that as the size of opening increase the ultimate load decreases while the deflection increases as expected.

As shown in Table 3; in the rectangular opening: a decrease percentage of (11.87%, 29.6%) in B9, and B2* respectively, compared to the normal arched beam without opening B2. in the circular opening: a decrease percentage of (9.4%, 27.5%) in B5, and B11 respectively, compared to the normal arched beam without opening B2.

5.3 Effect of Shape of Opening

The effect of aspect ratio of the opening on the load-deflection curve and ultimate load capacity in a circular arched beam with opening is studied. In this study different aspect ratios ranges from 1 to 3 for two different areas, as well as circular openings of the same areas at mid span of the beam are analyzed. Results show that the circular opening has the minimal deflection and the highest ultimate load among all opening shapes. As the aspect ratio of the opening decreases the load carrying capacity slightly increase. However, the aspect ratio has a negligible effect on the deflection and the ultimate load of arched beam as they share the same area of opening so the increase in

Table 4. The effect of shape of opening on ultimate load at different size

	Arched symbol	Shape of opening	Dimension (mm) $h_o * L_o$	Aspect ratio	Ultimate load (kN)	Decrease of ultimate load % compared to B2
Without opening	B2	NA	NA	NA	160	–
Area1 = 9375 mm^2	B4	Square	96.82 * 96.82	1	143	10.6%
	B5	Circle	D = 110	1	145	9.4%
	B6	Rectangle 1	79.06 * 118.59	1.5	142.5	10.9%
	B7	Rectangle 2	68.47 * 136.93	2	142.67	10.8%
	B8	Rectangle 3	61.24 * 153.09	2.5	141.33	11.7%
	B9	Rectangle 4	55.90 * 167.71	3	141	11.9%
Area2 = 22500 mm^2	B10	Square	150 * 150	1	115	28.1%
	B11	Circle	D = 169	1	116	27.5%
	B12	Rectangle 1	122.47 * 183.71	1.5	113.75	28.9%
	B13	Rectangle 2	106.07 * 212.13	2	113.375	29.1%
	B14	Rectangle 3	94.87 * 237.17	2.5	113	29.4%
	*B2**	Rectangle 4	86.60 * 259.81	3	112.658	29.6%

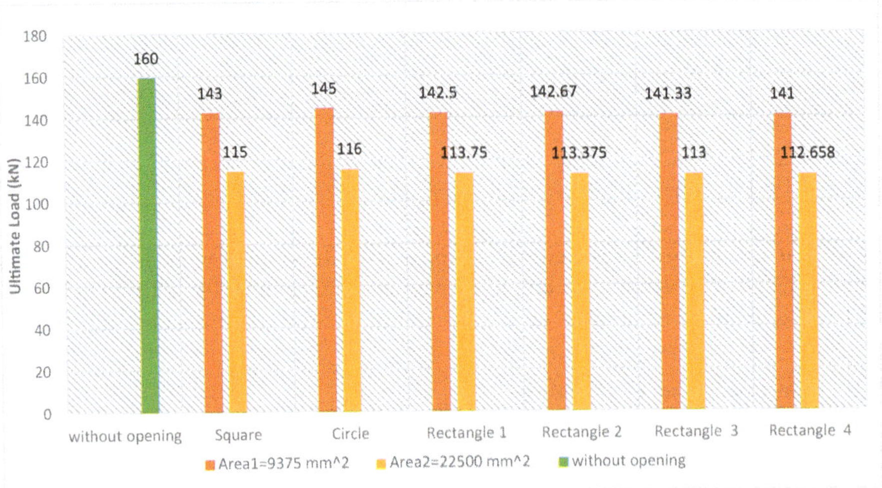

Fig. 16. The ultimate load at different size & shapes of opening

length of the opening compensate the decrease in its width, as shown in Table 4.

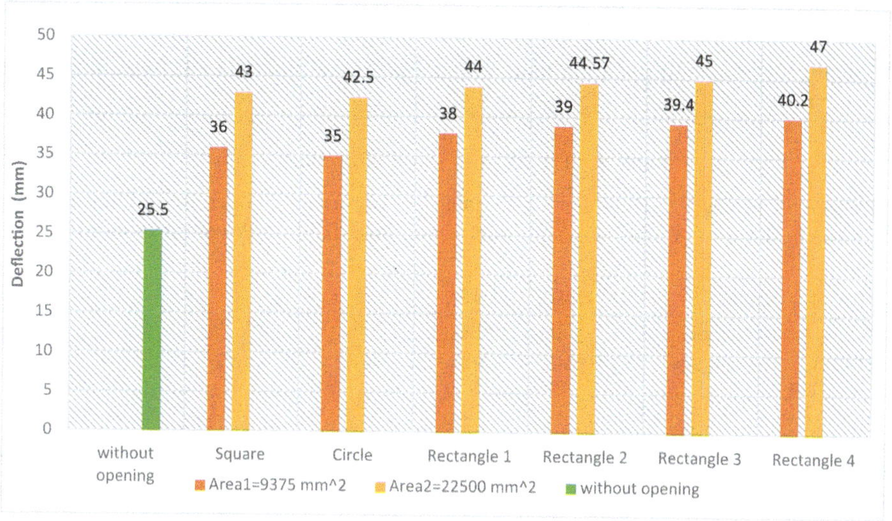

Fig. 17. Deflection at different size & shapes of opening

Bar chart in Figs. 16 and 17 compare between the effect of two different areas (area1 = 9375 mm^2 and area 2 = 22500 mm^2) for circular opening shape and various aspect ratios ranged from 1 to 3 rectangular openings on ultimate load and deflection respectively. Results show that the shape of the openings in the mid span position have a neglected effect on deflection and ultimate load. However, the circular opening has the minimal deflection and the maximum ultimate load.

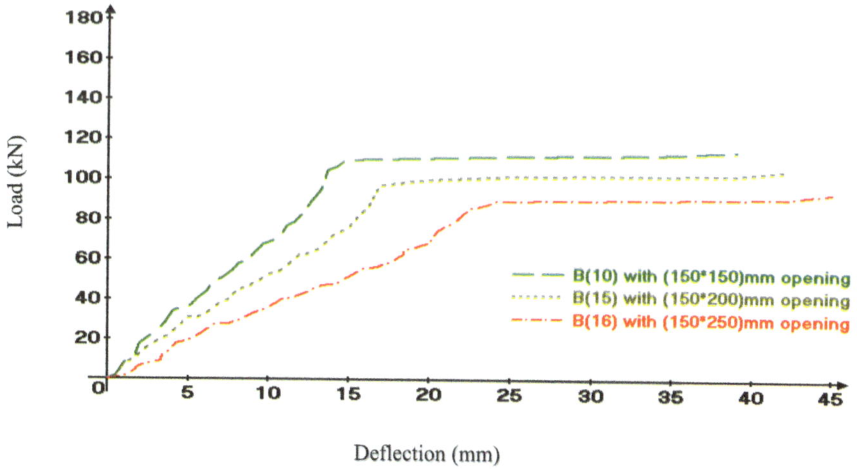

Fig. 18. Effect of opening length

5.4 Effect of Opening Length

Ultimate load capacity of a circular arched beam with different opening lengths are

Table 5. The effect of shape of opening length

Beam symbol	Opening shape	Dim. (mm) $h_o * L_o$	Opening area	Ultimate load (kN)	Decrease of ultimate load % compared to B10
B10	Square	150 * 150	22500	115	–
B15	Rectangle	150 * 200	30000	105	8.7%
B16	Rectangle	150 * 250	37500	94.3	18%

numerically investigated. The height of the opening is kept constant (150 mm) which equals 60% from the beams depth while its length is increased from 150 mm to 250 mm by increment of 50 mm. which equals a range from 8.6% to 14.3% of the beam span. Figure 18 shows the variation in load-deflection curve of the three cases.

It could be noticed that increasing in opening length leads to a decrees in ultimate load capacity of the beam due to the increasing in area and slenderness ratio as shown

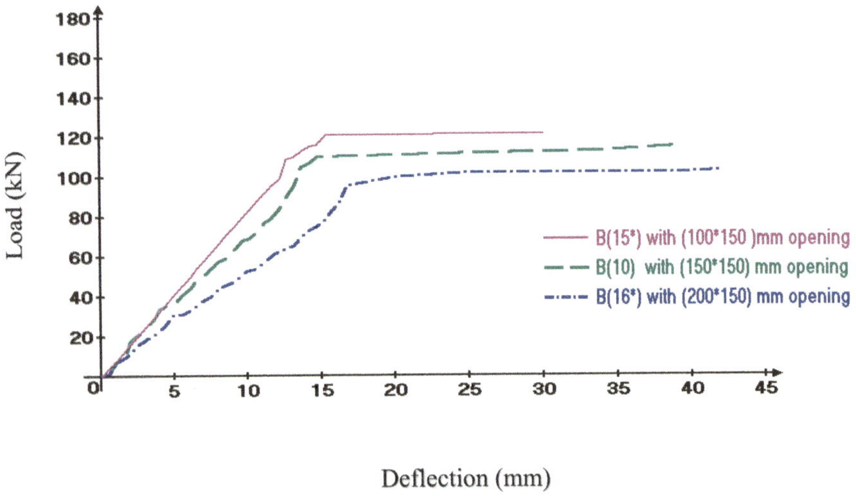

Fig. 19. The effect of opening height

in Table 5.

Table 6. The effect of shape of opening length

Beam symbol	Opening shape	Dim. (mm) $h_o * L_o$	Opening area	Ultimate load (kN)	Decrease of ultimate load % compared to B15*
B15*	Rectangle	100 * 150	30000	122	–
B10	Square	150 * 150	22500	115	5.7%
B16*	Rectangle	200 * 150	37500	103	15.6%

5.5 Effect of Opening Height

The effect of opening height in a circular arched beam at its mid span on ultimate load and deflection is studied. The length of the opening is kept constant (150 mm) which equals 8.6% from the beam span while its height is increased from 100 mm to 200 mm by increment of 50 mm, which equals a range from 40% to 80% of the beam depth. Figure 19 shows the variation in load-deflection curve of the three cases.

It is concluded that increasing the opening height with fixing the width of the opening leads to a decrease in ultimate load capacity of the beam, due to the reduction in cross section area of concrete, as shown in Table 6.

From the last two studies concerning the effect of height and length of the opening, and by comparing B15 and B16* which have the same area of opening and the same aspect ratio of 1.3334. However, the higher length in B15 is the length of the opening and in B16* the higher length is its height. Results reveal that B15 have slightly higher ultimate load than B16* due to higher concrete area in its cross section.

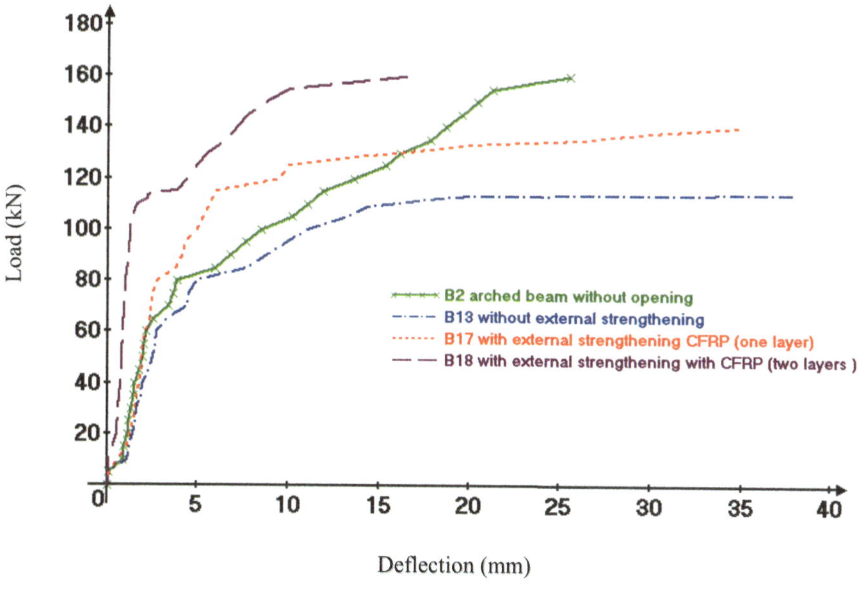

Fig. 20. The effect of number of layers on B13, B16, B17

5.6 Effect of Strengthening Number of CFRP Layers

In order to study the number of CFRP layers effect on ultimate load capacity of arched concrete beam with opening, one and two CFRP layers are compared with beam

Table 7. The effect of CFRP layers on ultimate load

Beam symbol	Number of layers of CFRP laminates	Opening $h_o * L_o$	Ultimate load (KN)	Increase of ultimate load % compared to B13
B2	NA	NA	160	28.8%
B13	NA	106.07 * 212.13	114	–
B17	One	106.07 * 212.13	140	18.6%
B18	Two	106.07 * 212.13	160	28.8%

without strengthened. The study carried out on circular arched beam with opening (106.06 * 212.13) mm^2 at the mid span of the beam. B13 without CFRP, B17 with one layer of CFRP and B18 with two layers of CFRP. The load-midspan deflection curves of all types are shown in Fig. 20. A significant difference in general response and ultimate load are recorded. There is a significant deformation happens after load 100

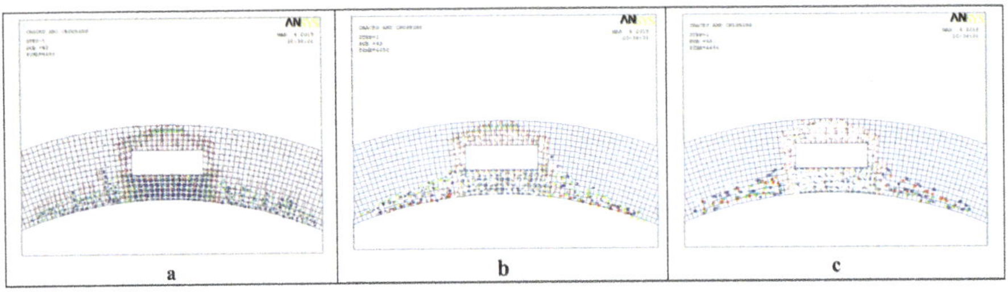

Fig. 21. The cracks pattern for beams B13, B17, B18 a. Cracks for B13 b. Cracks for B17 c. Cracks for B18

kN for all beams.

It can be concluded that when provide layers of CFRP laminates in arched beam.

A considerable increasing in the load capacity with percent increase of (18.6% and 28/8%), for B17 and B18 beams respectively compared to B13 beam, as shown in Table 7.

Figure 19 gives the clear idea of investigation & study of crack patterns in RC arched beam specimens (B13, B17, and B18) at load pattern 110 kN. Figure 21a. shows crack pattern for B13 without wrapping. The cracks are wide, considerable and

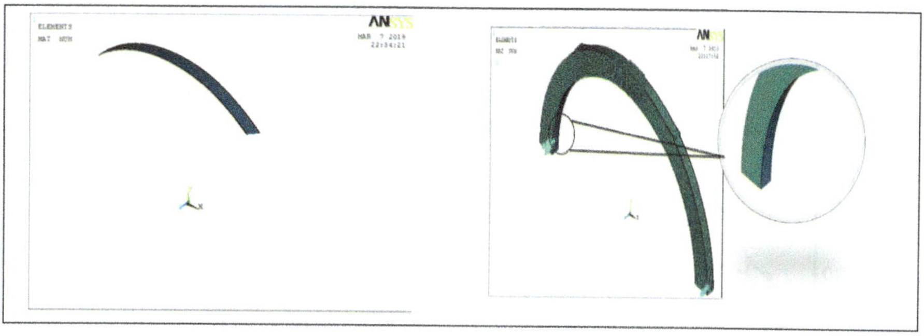

Fig. 22. Externally sheet of CFRP bonded at soffit of (B19)

become more uniform across the member. By adding one layer of CFRP represented in Fig. 21b. for B17 crack decreased significantly. In Fig. 21c. for B18 with two layers of CFRP the cracks percentage decreased more than B13 and B17.

Table 8. CFRP length development

Beam symbol	Opening size mm	CFRP length mm	Percentage of CFRP from beam length	Ultimate load kN	Increase of ultimate load % compared to B13
B13	106.07 * 212.13	0	0%	114	–
B19	106.07 * 212.13	106.5π	12.17%	130.3	14.3%
B20	106.07 * 212.13	150π	17.14%	132	15.8%
B21	106.07 * 212.13	250π	28.6%	134	17.5%
B22	106.07 * 212.13	290π	33.1%	134.370	17.9%
B23	106.07 * 212.13	875π	100%	135	18.4%

5.7 Effect of Externally Bonded CFRP Laminate at Beam Soffit

A CFRP laminate externally bonded at the soffit of beam is used along the lower chord of the arched beam span as shown in Fig. 22. To study the effect of CFRP laminate length on the ultimate load and deflection; different laminate lengths at soffit of beam are studied. Its length ranges from 213 mm which equals about 12% of beam length to 1750 mm which equals 100% of beam length. Beams are tested under two point loads at extrados (top) surface with applied load 180 KN.

Table 8 shows CFRP length development start from 0 mm to 1750 mm (full beam span) and its effect on ultimate load and deflection. Optimum one is found with length equal about 900 mm which equals to the mid span (B22). Increasing the length of CFRP more than this will have neglected effect on ultimate load and deflection.

Figure 23 shows the effect of soffit CFRP length on the deflection and ultimate load. As the length increase beneath the opening the deflection decreases and the ultimate load increase until certain limit (290π in this case study) more increase in the

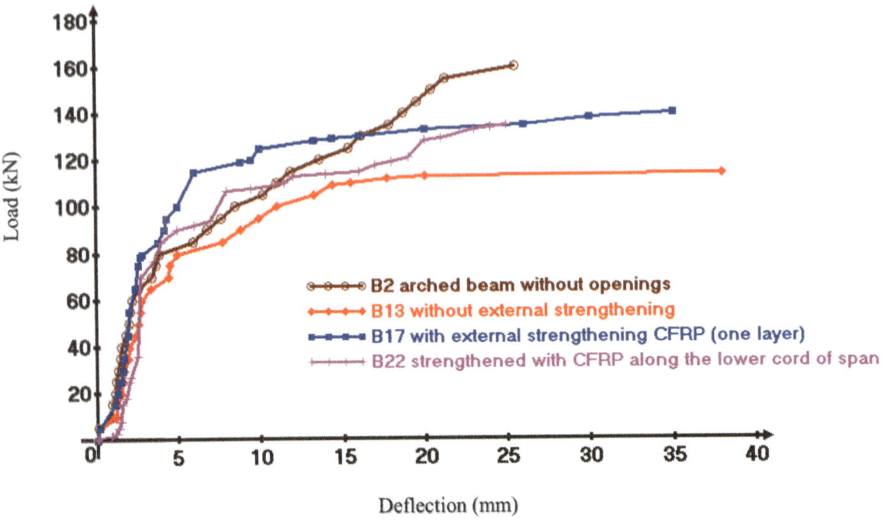

Fig. 23. Effect of externally boned soffit CFRP length on Deflection and Ultimate load

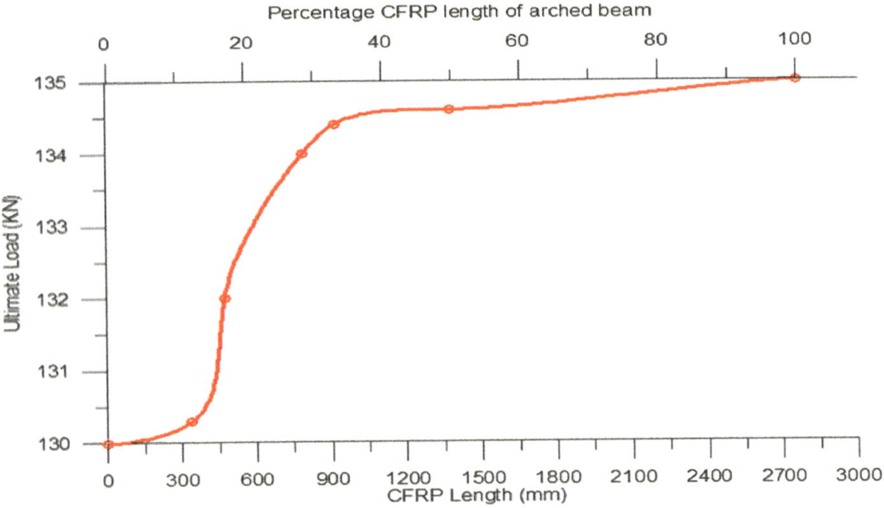

Fig. 24. Comparison between the different schemes of strengthen

soffit length will be in vain regarding its minimal effect on the deflection and ultimate load.

5.8 Comparison Between the Different Schemes of Strengthen

Strengthening process is highly affecting on the ultimate load capacity of R.C arched beams with openings using any strengthening type scheme. It can be divided into two

Table 9. Comparison between different schemes of strengthen

Beam symbol	Number of layers of CFRP laminates	Strengthening method	Ultimate load (KN)	Increase of ultimate load % compared to B13
B2	NA	NA	160	40.4%
B13	NA	NA	114	–
B17	One	Wrapping	140	22.8%
B22	One	Externally bonded to the beam soffit	134.37	17.9%

types: wrapping type as beam (B17) and externally bonded at soffit of the beam type as beam (B22).

Figure 24 Shows the results of the F.E. analysis results on load-deflection and ultimate load curves for the four arched beams (B2, B13, B17 and B22).

It could be conclude that the wrapping method has minimal deflection and the maximum ultimate load. Wrapping method increases the ultimate load by about 22.8% compared to beam without strengthen (B13). Externally bonded to the beam soffit method increases the ultimate load by about 17.9% compared to beam without strengthen (B13), as shown in Table 9.

6 Conclusion

The main conclusions observed from each phase of investigation for reinforced concrete arched beam with and without openings, internally strengthened by steel reinforcement or externally strengthened by CFRP laminates are as follows:

- For a fixed span increasing the curvature of beam rises the ultimate load capacity. Beam with H/L = 1 without opening has about 26% higher percentage of ultimate load than straight controlled beam without opening. Beam with H/L = 1 with opening has about 42.7% higher percentage of ultimate load than straight controlled beam with opening.
- As the area of opening in the mid span arched beams increases, the ultimate load decreases and the deflection increases.
- For the same area of opening, as the aspect ratio of the opening decreases, the load carrying capacity slightly increase. However, the effect of the aspect ratio on ultimate load could be neglected due to its small variation.
- Circular openings has minimal deflection and maximum ultimate load compared to rectangle opening with any aspect ratio for the same area.
- For fixed height of opening, increasing the opening length decreases the ultimate load capacity of the beam due to the increasing in area of opening and its slenderness ratio.
- For fixed length of opening, increasing the opening height decreases the ultimate load capacity of the beam, due to the reduction in cross section area of concrete.

- For the same area of opening and the same aspect ratio, the height of opening has slightly higher effect than its length on the ultimate load
- An increase in load capacity for arches wrapped by CFRP laminates one layer and two layers by about (18.6% and 28.8%) respectively when compared with that arch beam with opening without strengthen.
- As the length of CFRP length externally bonded at soffit of beam increase beneath the opening the deflection decreases and the ultimate load increase until certain limit. More increasing in the soffit length will be in-vain regarding its minimal effect on the deflection and ultimate load.
- Wrapping method gives better results than external bonded CFRP laminate at soffit of the beam for the ultimate load and deflection.

References

Al-Mutairee, H.M.K.: Effect of non-uniform distribution of longitudinal reinforcement on the behavior of reinforced concrete horizontally curved beams with fixed-ends. J. Univ. Babylon **21**(3), 826–838 (2013)

Ali, A.Y.: Three-dimensional nonlinear finite element analysis of reinforced concrete horizontally curved deep beams. J. Babylon Univ. **18**(1) (2010)

Ali, A., Hemzah, S.A.: Nonlinear analysis for behavior of RC horizontally semicircular curved beams with openings and strengthened by CFRP laminates. Int. J. Sci. Technol. Res. **3**(8), 136–145 (2014)

Ali, A.Y., Hamza, B.A.: Finite element analysis of RC Arches with openings strengthened by CFRP laminates. In: COMPLAS XIII: Proceedings of the XIII International Conference on Computational Plasticity: Fundamentals and Applications, pp. 495–506. CIMNE (2015)

Balamuralikrishnan, R., Antony, J.C.: Flexural behavior of RC beams strengthened with carbon fiber reinforced polymer (CFRP) fabrics. Open Civ. Eng. J. **3**, 102–109 (2009)

Guide, A.F.U.: Release 14.0, ANSYS. Inc., USA, November (2011)

Use of Granite Slurry in Masonry Manufacturing and Wall Construction

Basma Elsobky[1], Nada Attia[1], Omar Moustafa[1], Rafik Yanni[1], Mohamed Abdeldayem[1], Omar Elweleilly[1], Ezzat Fahmy[1], Mohamed Darwish[1,2(✉)], Mohamed Rashwan[1], and Mohamed Nagib Abouzeid[1]

[1] American University in Cairo, Cairo, Egypt
mdarwish@aucegypt.edu
[2] Ahram Canadian University, Giza, Egypt

Abstract. The waste powder produced during granite manufacturing has been a source of significant pollution and health hazards. The use of this waste powder in concrete has been studied by several researchers. The current study involves partial replacement of the fine aggregates in the mix of concrete masonry by granite waste powder. The Brazilian split test and the slant test were performed to check the bond with mortar. The mix was used to construct walls subjected to out-of-plane loading and compared to a control group. The initial results indicate that incorporating such waste in walls can be one promising venue for its utilization in the construction industry.

1 Introduction

Egypt has got huge stores of amazing granite and stone. World Stone generation achieved the pinnacle of somewhere in the range of 75 million tons (or 820 million m^2 proportionate), net of quarry waste (El-Sayed 2016). The official generation figures of Egypt are striking; yet, the genuine creation is extensively higher than the level demonstrated by the official insights and possibly past the levels assessed over the span of the examination featured. The estimations in light of the data recovered through neighborhood appraisal credited to Egypt: a quarry generation of around 3.2 million tons and more than 25 unique sorts of Egyptian stone (El-Sayed 2016). This shows the nation lies among the main 8 world makers of crude material. The normal yearly rate of increment has achieved 8.8% since 2002 (Allam 2014).

Concerning quarry creation and crude fare, Egypt is positioned the fourth separately with an offer of 4.3% and 6.6% of aggregate world market of Granite and rock. This implies Egypt can be viewed as the seventh exporter on the planet, as far as volume, after China, India, Italy, Spain, Turkey and Brazil. During the process of granite manufacturing, a massive amount of water is used while cutting the granite piles with diamond blades to cool off the process. The result is water absorbing dust and granite slurry powder without any proper way to recycle it or getting rid of it. If the water is disposed on the ground, the slurry powder dries, and it causes air pollution.

H. Rodrigues et al. (Eds.): GeoMEast 2019, SUCI, pp. 39–47, 2020.
https://doi.org/10.1007/978-3-030-34216-6_3

The main issue that water use in the granite manufacturing process is essential to cool down the blades.

Several researchers have worked on utilizing the use of this slurry powder byproduct in cement bricks. One of these works was done by Hamza et al. (2011) who produced cement bricks with partial replacement of the aggregates with the slurry. The partial replacement was done for increments of 10%, 20%, 30% and 40% of the aggregates. The bricks were tested for physical and mechanical properties and the test results revealed that the recycled products have physical and mechanical properties that qualify them for use in the building sector, where all cement brick samples tested in this study comply with the Egyptian code requirement for structural bricks, with granite slurry having a positive effect on cement brick samples that reach its optimum at 10% slurry incorporation (Hamza et al. 2011).

Similarly, other researchers such as (Allam and Bakhoum 2014), (Kumara 2015) and (Dhanapandian and Shanthib 2009) did similar works to that done by Hamza et al. (2011). However, none of these researchers narrowed down the window of the study of varying the percentage slurry than increments of 10%. Also, none of them studied the durability of the produced bricks or their performance under axial loads in the walls. These issues are important as they will affect the decision of using such products in various conditions of loading and in various environments.

The research performed by Attia et al. (2018) covered the variation in usage of the percentage of slurry ranging from 0% to 25% slurry in replacement of aggregates and tested for compressive strength. The units having 7.5% slurry were found to have the highest compressive strength. Also, the durability of the bricks was studied by immersion in salt solutions in addition to regular fresh concrete tests. Finally, the performance of brick prisms under compression was studied and showed that the prisms involving bricks having 7.5% granite slurry had sufficiently high compressive strength (Attia et al. 2018).

The current research presented in this paper is a continuation of the work performed by Attia et al. (2018). The target of the current research is to assess the performance of the bricks having the 7.5% granite slurry within the walls. This was performed by performing the slant test and the Brazilian indirect split test in order to assess the bond between the bricks. Furthermore, wall samples were constructed using the developed bricks and tested till failure under out-of-plane loading.

2 Experimental Program

The experimental work that was conducted by Attia et al. (2018) involved slurry replacement of fine aggregate at different increments of ranging from 0% to 25%. The highest three mixes were chosen based on their compressive strength. As the best mix involved 7.5% slurry, this same mix was used within the current study. The bond strength is assessed by performing the slant test and the Brazilian split test. Finally, the highest mix is used in casting 3 walls from the developed mix and 1 control wall so the total becomes 4 walls.

2.1 Material Properties

The materials used within the study were as follows:

- ***Cement:*** Type I Ordinary Portland Cement Concrete with specific gravity of 3.15.
- ***Fine Aggregates:*** Natural sand will be used. The sand had absorption of 0.4%, S.S. D specific gravity of 2.55 and a fineness modulus of 2.85.
- ***Coarse Aggregates:*** Well-graded crushed dolomite was used as coarse aggregates. The aggregates had absorption of 1.8%, MNA of 38 mm and an S.S.D specific gravity of 2.57.
- ***Water:*** Ordinary municipal tap water was in washing the aggregates as well as the production and curing of the concrete mixtures.
- **Granite slurry:** A Granite slurry powder was used, which was obtained from a plant from Alexandria from granite quarried from Aswan.

2.2 Mix Design

The 7.5% slurry mix involved 493 kg of cement per m^3, 50 kg of granite slurry per m^3, 612 kg of fine aggregates per m^3, 998 kg of coarse aggregates per m^3 and 185 kg of water per m^3.

2.3 Tests Performed

2.3.1 Brazilian Indirect Split Test

One of the tests performed to measure the bond between the mortar and the developed units was by conducting the indirect split test which is also known as "Brazilian test". This test was conducted by preparing cylinder samples and splitting them into two halves and then sticking them again with a cement and analyze the value and the place at which the failure happens when exposed to load. This was performed for three control samples involving 0% slurry and three samples involving 7.5% slurry. The procedure is as follows:

1. The mix was prepared and then poured inside the cylindrical shape to come up with a cylinder sample.
2. The sample were then left to dry for 1 day.
3. The sample entered the curing room for 1 week.
4. After that, the sample was then cut into two halves and then, 10 mm of mortar is placed (same mortar mix cubes used by Attia et al. (2018)) between the two surfaces of the cylinder and left for 14 days as shown in Fig. 1.
5. The sample was then tested for the compressive strength by applying perpendicular load to the cylinder, the place at which the cracks begin could be observed as well as the value of the compressive load.

2.3.2 Slant Test

One of the tests performed to measure the bond between the developed units was the slant shear test. This test was conducted according to ASTM C882 M by preparing brick samples and splitting them into two halves at an angle of 26.6° with the axial

Fig. 1. The Brazilian test samples ready to conduct the test

direction of the cylinder to act in the plane of shear of the sample when subjecting the prepared sample to compression in the axial direction of the sample. This compression is applied 7 days after applying the layer of mortar between the two parts of the sawed sample. The procedure was performed according to ASTM C882 M standards however the only two deviations were that this standard was specified for epoxy resins however ordinary Portland cement mortar was used in this study and the samples were brick samples not cylinders as specified by the standard (Fig. 2).

Fig. 2. The out-of-plane test setup.

2.3.3 Out-of-Plane Load Test

One of the main necessary items within this research is to analyze the structural performance of these bricks when built as a wall. Walls having a height of 1 m, a width of 1 m and a thickness of 120 mm were constructed. An out of plane test on three walls build up from the 7.5% slurry bricks was performed. This was also done for a control samples involving no slurry. The load was applied perpendicular to the wall while standing, to imitate the wind loading scenario in a displacement control mode of loading. The procedure is as follows:

1. Steel channels were designed taking into consideration the wall dimensions to tighten the wall from its top and bottom sides while applying the load.
2. The steel channels are connected to each other using two long screws that hold the two steel channels to each other.
3. The wall is then placed in position and the load cell starts applying the load.
4. The load applied on each wall has at a rate of 6 mm/min.

3 Results

3.1 Brazilian (Indirect-Split) Test

The results for the 0% and 7.5% slurry are shown in Table 1. The average ultimate load of the samples involving 7.5% slurry was nearly four times that of their control counterparts. Hence, the bond strength of the samples involving 7.5% is considered to be promising which will be further investigated in the coming tests.

Table 1. Results of the Brazilian (indirect-split) test.

Slurry percentage	Average ultimate load (kN)
0%	22.7
7.5%	84.7

3.2 Slant Test

In order to confirm the strength of the sample 7.5% of the granite waste, we have conducted the Slant test. The experiment shows the bond between the mortar and the brick, as its two samples connected with mortar at an angle. The results showed that this bond between the mortar and the bricks are so high. The failure occurred in the mortar at compressive strength of 49.5 KN in which the 7.5% slurry was used.

3.3 Out-of-Plane Load Test

Within all the tested specimen, failure occurred in the mortar having a horizontal crack propagating along the wall causing failure as showed in Fig. 3. This occurred in the 4 samples including the control sample.

Fig. 3. One of the out-of-plane tested 7.5% slurry samples.

The load-deflection curve of the 1st 7.5% sample shown in Fig. 4 shows that an ultimate load of 8.5 kN is reached at a 30 mm displacement and failure happens at a displacement of 44 mm. The load-deflection curve of the 2nd 7.5% sample shown in Fig. 5 shows that an ultimate load of 7.8 kN is reached at a 20 mm displacement and failure happens at a displacement of 34 mm. The load-deflection curve of the 3rd 7.5% sample shown in Fig. 6 shows that an ultimate load of 7.1 kN is reached at a 26 mm displacement and failure happens at a displacement of 34 mm.

On the other hand, the load-deflection curve of the control sample shown in Fig. 4 shows that an ultimate load of 5 kN is reached and failure happens at a displacement of 34 mm. Hence, when comparing the 7.5% wall samples to the control sample, the samples had an average ultimate load that is almost 1.5 times the ultimate load that the control wall could withstand. That is a very good and crucial indicator of the structural performance of these bricks when built up as a whole unit. These results are in agreement with the large values of bond strengths reached when performing the Brazilian in-direct split test and the slant test which mean that it is expected to see a significant increase in the bond between the units within a wall as shown in the out-of-plane test (Fig. 7).

4 Conclusions and Recommendations

In lieu of the results the following conclusions could be withdrawn:

- Manufacturing of bricks with granite slurry contribute to both environmental as well as technical merits as the incorporation of slurry into the bricks commute to the enhancement of mechanical properties.

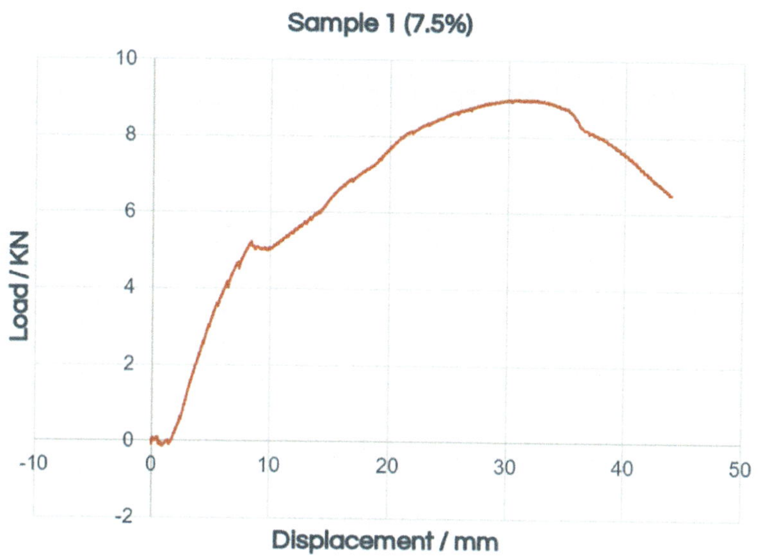

Fig. 4. The out of plane load-deflection curve of the 1st 7.5% slurry sample

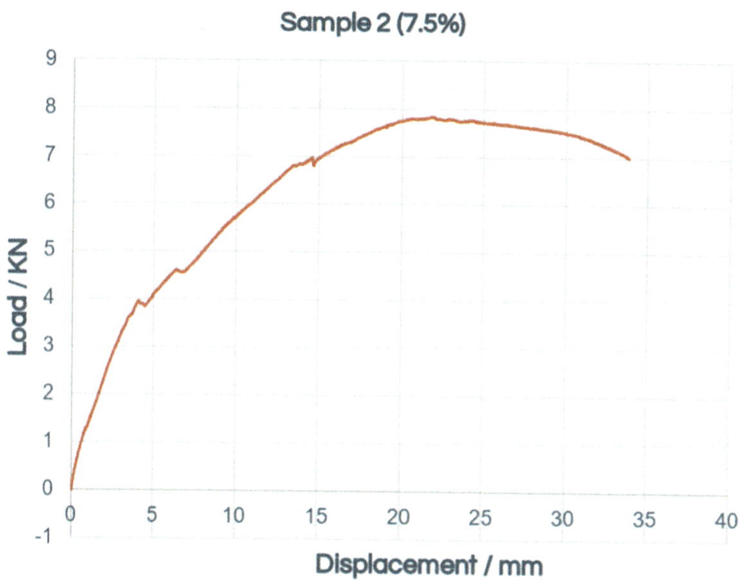

Fig. 5. The out of plane load-deflection curve of the 2nd 7.5% slurry sample

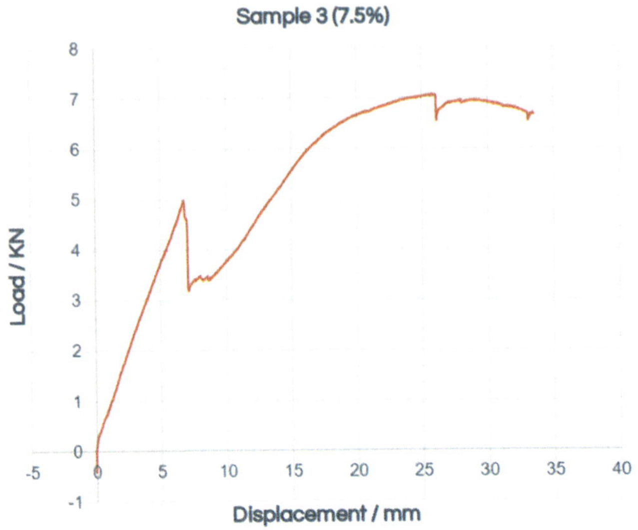

Fig. 6. The out of plane load-deflection curve of the 3rd 7.5% slurry sample

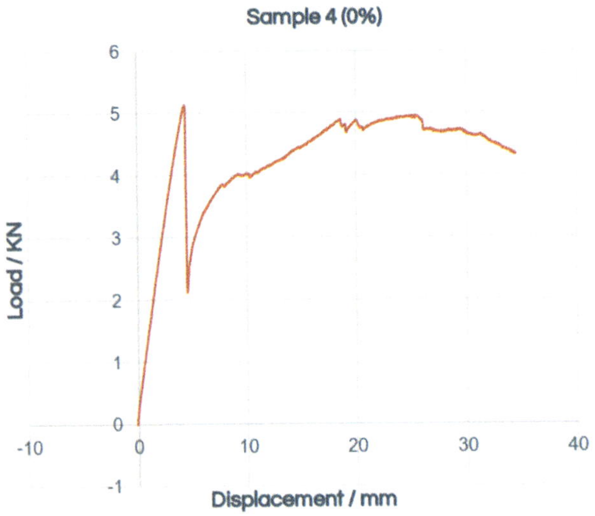

Fig. 7. The out of plane load-deflection curve of the 0% slurry (control sample)

- The bond between the masonry units involving 7.5% slurry has been found to be significantly higher than that of the samples having no slurry in them.
- For the walls, it was concluded that these bricks are able to withstand a high out-of-plane load which is about 50% more than the load that the control wall can withstand.
- The failure pattern that occurred in the mortar gives a very good indication of the structural performance of the 7.5% wall.

In lieu of these conclusions the following actions are recommended:

- Study the feasibility of using the slurry in masonry manufacturing on an industrial scale.
- Study the effect of varying the percentage of slurry when using commercially used mixtures having lower cement content.

Acknowledgements. The authors would like to acknowledge the efforts of the teaching assistants, lab engineers and lab technicians in the construction engineering labs at the American University in Cairo for their extensive help and assistance during performing the work.

References

Allam, M.E., Bakhoum, E.S.: Re-use of granite sludge in producing green concrete. ARPN J. Eng. Appl. Sci. **9**(12), 2731–2737 (2014)

Attia, N., Moustafa, O., Abdeldayem, M., Yanni, R., Elsobky, B., Elweleily, O., Darwish, M., Rashwan, M., Abouzeid, M., Fahmy, E.: Use of granite manufacturing waste in concrete masonry production. In: 7th International CSCE Materials Specialty Conference, Fredricton NB Canada, 14–17 June 2018

El-Sayed, H.A.: Characteristics of the marble processing powder waste at Shaq El-Thoaban industrial area, Egypt, and its suitability for cement manufacture. HBRC J. **14**(2), 171–179 (2016)

Kumara, V.: Experimental study on granite sludge blocks. Int. J. Inform. Futuristic Res. **2**(10), 3773–3781 (2015)

Mehta, J., Pitroda, J.: Effective utilization of granite powder as replacement of marble powder in marble brick, vol. 2, no. 3, May 2016. https://doi.org/10.15438/rr.4.4.22

Hamza, R.A., El-Haggar, S., Khedr, S.: Marble and granite waste: characterization and utilization in concrete bricks. Int. J. Biosci. Biochem. Bioinform. **1**(4), 286–291 (2011)

Dhanapandian, S., Shanthib, O.M.: Utilization of marble and granite wastes in brick products. J. Ind. Pollut. Control **25**(2), 145–150 (2009)

Standard Test Methods for Sampling and Testing Concrete Masonry Units and Related Units, ASTM C140/C140 M – 17a

Standard Test Methods for Conducting Strength Tests of Masonry Wall Panels, ASTM C1717 – 12

Standard Test Method for Compressive Strength of Masonry Prisms, ASTM C1314 – 16

Standard Test Method for Bond Strength of Epoxy-Resin Systems Used With Concrete By Slant Shear, ASTM C882 / C882 M - 13a

High Performance Materials for Concrete Bridge Construction

George Morcous[1(✉)], Marc Maguire[1(✉)], and Maher K. Tadros[2]

[1] Durham School of Architectural Engineering and Construction, University of Nebraska-Lincoln, Omaha, NE, USA
gmorcous2@unl.edu, mmaguire75@gmail.com
[2] e-Construct.USA LLC, Omaha, NE, USA
maher.tadros@econstruct.us

Abstract. Precast/prestressed concrete is commonly used in short and medium span bridge construction because its assured quality, inherent durability, and optimized use of materials satisfy the needs of bridge owners. Recent research projects conducted at the University of Nebraska – Lincoln (UNL) focused on using high performance materials in precast/prestressed concrete bridge girders and deck panels that significantly improve their durability, economics, and speed of construction. This paper discusses the use of High Strength Self-Consolidated Concrete (HSSCC), 18 mm diameter prestressing strands, and Grade 550 high strength welded-wire reinforcement (WWR) to achieve simplicity of production and erection and maximize span-to-depth ratio while being economical. Precast/prestressed concrete girders with span-to-depth ratio of over 30 were developed and tested. The main advantages of these girder are the ease of production, speed of construction, and enhanced durability. A summary of specimen design, production, and tested is presented.

1 Introduction

New high performance materials are constantly evolving to produce stronger, more durable and economical structures. Examples are: ultra-high performance concrete, high strength steel, large diameter strands, etc. The challenge is how to efficiently combine these materials to achieve design and construction requirements while being cost effective. The objective of this project is to utilize the advantages of High Strength Self-Consolidated Concrete (HSSCC), 18 mm diameter seven-wire prestressing strands, and Grade 550 welded wire reinforcement (WWR) while using existing precast concrete sections to develop a bridge superstructure system that minimizes structural depth, construction duration, and fabrication and construction costs.

Precast concrete producers are slowly implementing self-consolidating concrete (SCC), large diameter prestressing strands, and WWR in several applications to save labor cost and improve product quality. SCC eliminates the need for external/internal vibration and speed up concrete placement. The 18 mm diameter strands results in 92% increase in strand area over the most common 15 mm diameter strands. WWR results in cost savings due to its strength and ease of placement. By combining these high

H. Rodrigues et al. (Eds.): GeoMEast 2019, SUCI, pp. 48–63, 2020.
https://doi.org/10.1007/978-3-030-34216-6_4

performance materials, a more economical, shallower, stronger and durable concrete product can be readily available for bridge construction.

The research team selected standard PCI "heavy" double tee to develop efficient and practical precast/prestressed concrete girders for short and medium span bridges. Double tees, most often used in building/garage construction, are very easy to produce and are universally available. Double tee forms are often self-stressing and easily stripped. This makes them ideal for fast paced and economical fabrication. Efficiently using concrete in the double tee form, creative formwork blocking can produce significantly smaller depths to minimize concrete costs.

Figure 1 presents a possible block-out pattern, which was used in the present study. This practice will reduce concrete, which will reduce direct material costs, as well as weight, opening up bidding to construction crews with smaller cranes. Often, double tee girders will have all of their strands draped at one or two points such that all strands are touching between depression points. This practice dramatically increases strand eccentricity near midspan.

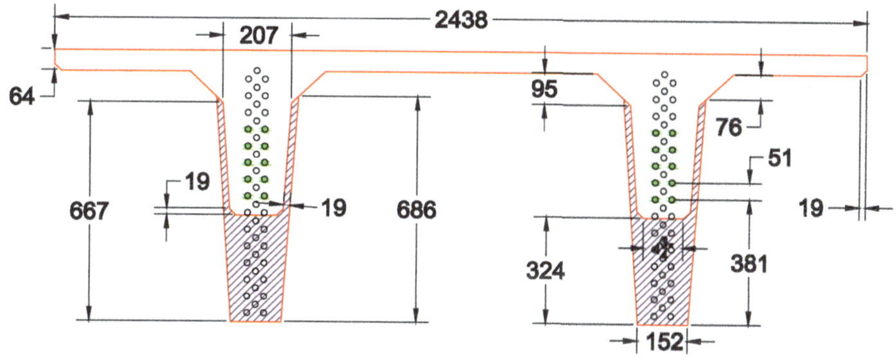

Fig. 1. PCI standard heavy double tee with specimen blockouts

During construction, double tee girders can be placed side-by-side, eliminating deck formwork for a cast-in-place (CIP) deck. Additionally, the wide, flat, top flange will promote worker safety. By using innovative transverse connections, the double tee girder would make an ideal candidate for accelerated bridge construction. Overall, these properties make the double tee form an ideal candidate for a high performance bridge system, while using existing PCI standard formwork.

2 Testing Scope and Setup

The 18 mm diameter strands and HSSCC are incremental improvements upon conventional materials. As such, it is important to study their behavior prior to full scale implementation. The HSSCC selected for the double tee system has several challenges: First, the very high strength and exceptional bond characteristics create smooth cracks that fracture through the aggregate. It was a concern that these cracks would limit aggregate interlock and therefore cause shear calculations to over-predict the concrete capacity. Second, upon placement, the HSSCC develops a thick skin. This skin prevents raking or surface roughening and results in a glassy surface. If bridge designers desire a concrete deck or topping, the horizontal shear connection is important. Third, there has been limited research involving 18 mm diameter strands and the resulting transfer and development lengths (Morcous et al. 2011a). Therefore, it will be necessary to determine if the large diameter strands can maintain appropriate bond under harped conditions in the HSSCC. Harping the tendons increases prestressing eccentricity, but the harping technique places the 18 mm diameter strands at 18 mm vertical centers at midspan. This represents the first time a precast producer has harped 18 mm diameter strands. It is important to validate this technique for the bridge system.

To study and validate the proposed bridge system, referred to as the Bridge Double Tee (BDT) system, a full-scale BDT girder was fabricated by Coreslab Structures, Inc., Omaha, Nebraska. Six load tests, performed on two full-scale prestressed concrete single tee bridge girders cast in the standard double tee form, made up the testing program. As outlined above, the main objectives were:

- Investigate the harped 18 mm diameter strand and HSSCC flexural capacity;
- Evaluate the shear transfer between the HSSCC girder and a cast-in-place (CIP) topping;
- Compare the transfer and development length of 18 mm diameter strands in HSSCC to design code values.

3 Girder Design and Fabrication

Using the 2008 American Association of State and Highway Transportation Officials (AASHTO) Load and Resistance Factor Design (LRFD) specifications, the research team designed a 15.5 m long, simple-span double tee bridge girder. The standard double tee form was blocked out to reduce stem thickness and depth, as shown in Fig. 1. To increase testing opportunities, the double tee was divided into two single tee girders.

The resulting precast cross-section was 603 mm deep with the 102 mm CIP composite deck. Girder concrete was specified to be 83 MPa at release and 103 MPa at 28 days. Based on prior experience (Morcous and Tadros 2009), the HSSCC would significantly outperform these expectations. Because the 18 mm prestressing strands exceeded the anchorage hole diameters, each hole was enlarged from 13 mm to 18 mm diameter. The strands, tensioned to $0.60f_{pu}$ to accommodate form capacity, were harped at 0.4L.

The BDT cross-section is presented in Fig. 2 with the left stem showing midspan details and the right stem with girder end details. The girder cross-section, shown in Fig. 2, shows the harping pattern of the 10 Grade 1860 low-relaxation 18 mm diameter strands and reinforcement per stem. Grade 550 MD71@152 mm WWR provided end zone confinement and shear reinforcement throughout the beam with MD52 cross wires to aid anchorage. Three 19 mm headed coil rods, welded to the steel base plate, reinforced the end zone as seen in Fig. 2, on the right stem. Two mats of 305 × 203 - MD129 × MD52 were used for deck reinforcement.

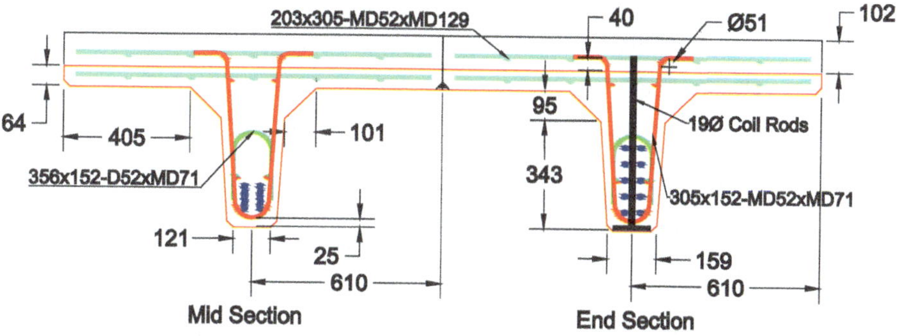

Fig. 2. Test specimen cross-section, left stem: midspan detail, right stem: end detail

Figures 3 and 4 present the girders during fabrication in the standard double tee form with the WWR in place. To facilitate bottom WWR mat installation, the shear reinforcement ends remained vertical during casting. Prior to deck placement, the reinforcement was bent according to Fig. 2. The confinement WWR has longitudinal bars that set on top of the longitudinal bars on the vertical shear WWR to confine the prestressing strands. Confinement was provided throughout the span, as is typical for Nebraska bridge girder fabrication. Figure 3 shows the strand pushdown location along with the protruding shear reinforcement in the self-stressing double tee form. The self-consolidating HSSCC easily filled the form, as shown in Fig. 4, and vibration was not necessary. Each single tee girder was filled using a single concrete batch.

Fig. 3. BDT girders prior to concrete placement

Fig. 4. HSSCC placed in double tee form

4 Girder Instrumentation and Test Setup

To determine the transfer length, each end of the BDT Girders was instrumented with 16 detachable mechanical (DEMEC) strain gauges, placed approximately 100 mm apart. The 16 DEMEC gauges allow for 14 surface strain readings along the girder ends. Because the form makes the stems inaccessible, DEMEC strain gauges were located along the top flange centerline rather than at the centroid of the strands. Strain readings were taken immediately prior to release, 30 min after release and after 14 days.

Figure 5 shows the testing plan and instrumentation layout for both girders. Each single tee beam underwent three tests. The first test placed the load at the AASHTO LRFD prescribed development length to verify code prediction. The second test was a midspan flexural test taken to failure. The final test was a shear test where the load was placed 1676 mm from the support. For each test, the deflection was measured using string potentiometers (S-POTs) located directly under the loading points. Linear potentiometers (L-POTs) measured strand end-slip nearest to the load for all tests. Electrical resistance strain gauges (ERSGs) monitored the difference in strain between the CIP and precast (PC) girder for each test 610 mm from the load. Each ESRG was oriented horizontally and located 6.5 mm vertically from the interface and 13 mm away from each other.

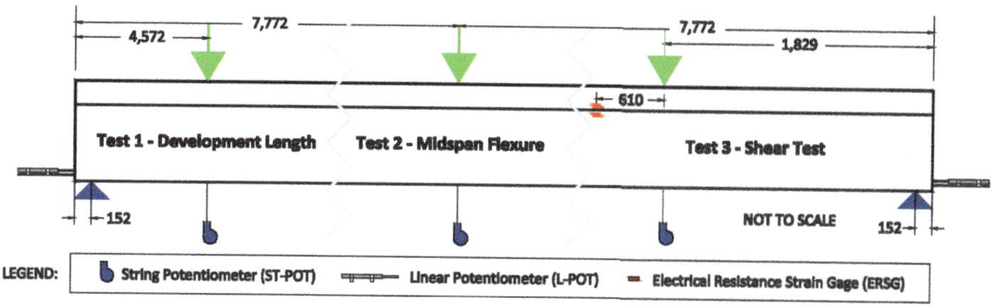

Fig. 5. Instrumentation plan for BDT

5 Test Results and Discussion

5.1 Material Properties

Figure 6 plots the precast concrete girder compressive strength versus age. Both girders exceeded the 103 MPa design cylinder design strength at 28 days. However, a minor steam system malfunction caused low curing temperatures, which resulted in slow strength gain and the design release strength was not reached at 18 h as planned.

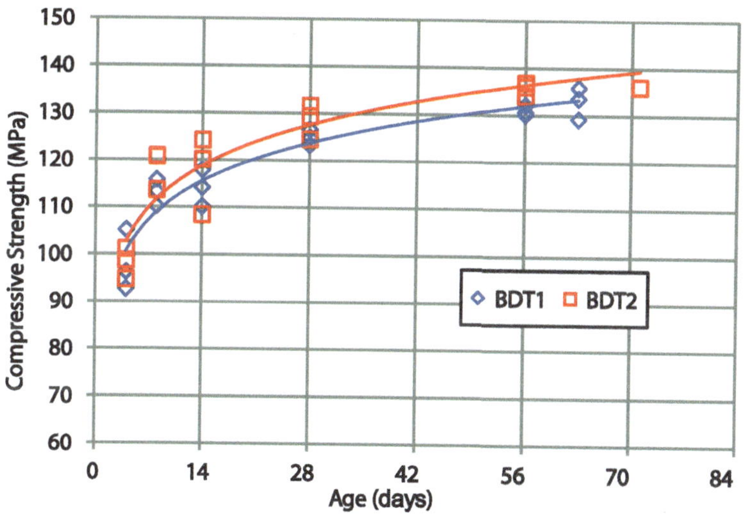

Fig. 6. Compressive strength versus age plot for BDT concrete

Stress-strain relationships for the BDT concrete were determined at release, 28 days and the day of load testing. Figure 7 compares the stress-strain relationships at different ages for the girder and deck concretes. Notice the linear relationship with minimal stiffness loss near maximum strength for the more mature concretes. For brevity, more on these relationships is presented in detail elsewhere (Maguire 2009). Table 1 summarizes girder and deck concrete measured values for compressive strength (ASTM

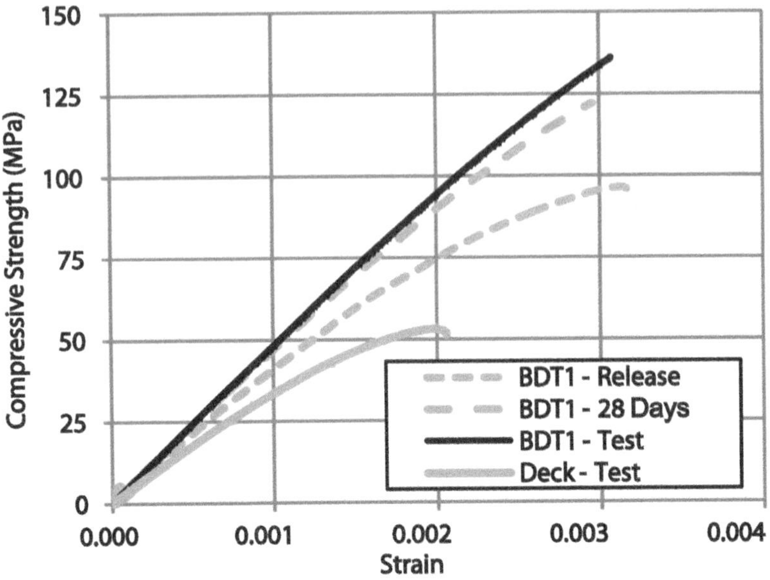

Fig. 7. Concrete stress versus strain comparison for BDT1

Table 1. Measured concrete mechanical properties

	BDT1	BDT2	Deck
28-day compressive strength (Mpa)	120	121	–
Final compressive strength (Mpa)	136	136	54
28-day MOE (Mpa)	46,036	44,954	–
Day of testing MOE (Mpa)	47,988	48,125	35,480
28-day poisson ratio	0.224	0.224	–
28-day MOR (MPa)	8.05	7.50	–
28-day splitting strength (MPa)	6.95	6.87	–

C39), Modulus of Elasticity (MOE, from strain gauges), Poisson's Ratio (from strain gauges), Modulus of Rupture (MOR, ASTM C78 2006) and splitting tensile strength (ASTM C496) at 28 days and girder test days.

Figure 8 shows stress-strain relationships for the six tests on 18 mm diameter seven-wire strands used for the girders in this study. The stress-strain curves had little variability and are nearly indistinguishable from one another. Testing procedure details are presented elsewhere (Maguire 2009). Table 2 presents tabulated test results and compares to minimum standard. This strand did not meet the ASTM A416 minimum load at 1% strain criteria and was slightly under the recommended MOE. This was not considered detrimental to the study.

5.2 Transfer Length

The literature suggests transfer length expands 10% to 20% over time (Barnes et al. 2003), with the majority of the extension coming in the first 14 days (Carrol et al. 2008). Therefore, these readings were considered the initial and final transfer lengths. The transfer lengths were calculated with a modified 95% Average Maximum Strain (AMS) method presented in Carrol et al. (2008). The transfer length values determined from the girder ends (north or south), with accompanying ACI 318 (2008) and AASHTO LRFD predictions, are tabulated in Table 3. The harped 18 mm diameter strands at 51 mm by 51 mm spacing in HSSCC present transfer lengths significantly below the predicted code values.

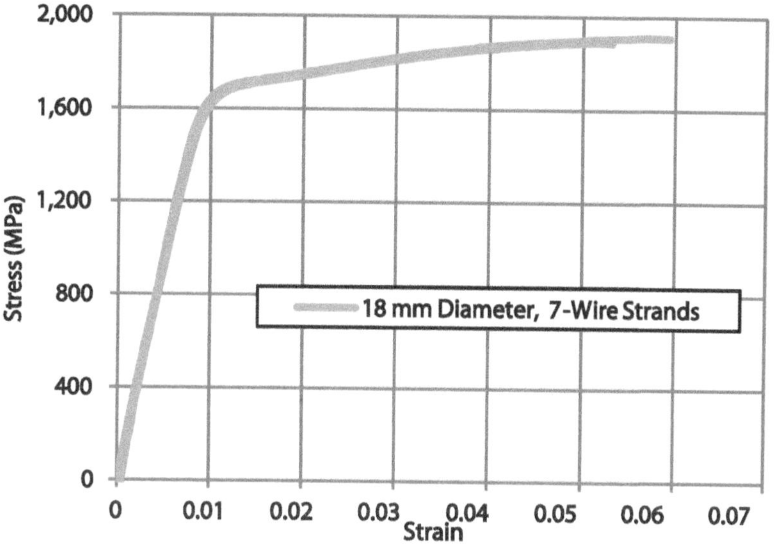

Fig. 8. Stress-strain relationships for 18 mm diameter seven-wire strand

Table 2. 18 mm strand tests compared to ASTM A416

Strand number	Load at 1%, kN	Peak load, kN	MOE, MPa
1	275	359	188,944
2	314	363	187,117
3	317	355	195,735
4	309	361	197,031
5	301	357	212,290
6	289	358	185,469
Average	**301**	**359**	**194,432**
Nominal minimum	*318*	*353*	*196,500*

5.3 Development Length Testing

Figure 9 presents the load deflection curves for Girders 1 and 2. During this test, the girders were not loaded to failure, due to the planned three-test regimen for each girder. However, the ultimate predicted applied load of 423 kN was applied at the development length. Ultimate load was predicted using strain compatibility with measured material properties. Bottom strand stress estimation, under this loading, was 1806 MPa based on strain compatibility. There was no significant slippage measured by the L-POTs on either girder throughout the loading. This indicates that the AASHTO LRFD prescribed development length of 4572 mm is conservative for 18 mm strands in HSSCC tensioned to $0.6f_{pu}$.

Table 3. Transfer length comparison

	At release, mm	At 14 days, mm	ACI $(f_{se}/3)d_b$, mm	AASHTO $60d_b$, mm
BDT1–S	445	536	841	1067
BDT1–N	518	462		
BDT2–S	368	447		
BDT2–N	345	429		
Average	419	470		

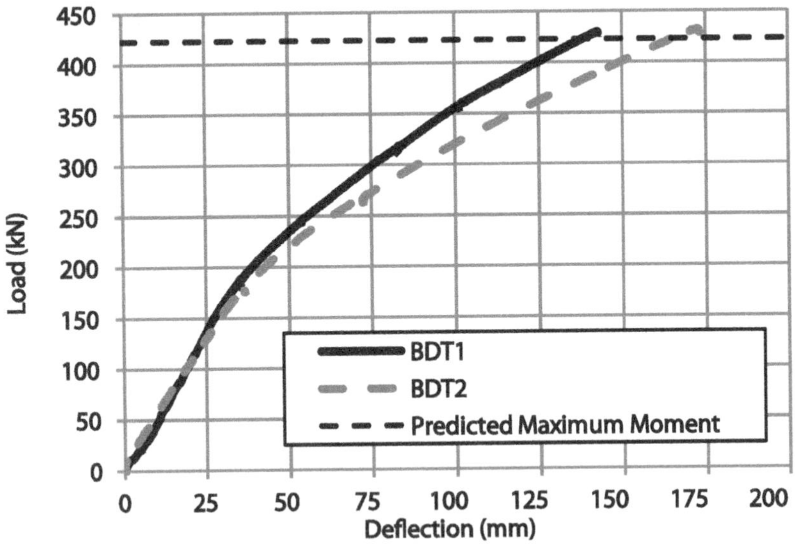

Fig. 9. Load versus deflection plot for development length tests

5.4 Ultimate Flexure Testing

Figure 10 shows the final deflected shape for BDT1. Deflection was very dramatic for both girders, as presented in the load versus deflection plots in Fig. 11. Both girders failed at approximately 405 kN and no significant slip was measured on either test. The predicted capacity was 378 kN at midspan, which is represented as a dashed line in Fig. 11. Both girders showed similar load versus deflection behavior throughout the test. This test indicates strength design is sufficient to predict the capacity of composite, HSSCC bridge girders with harped 18 mm diameter prestressing strands.

Fig. 10. BDT1 flexure test, final deflected shape

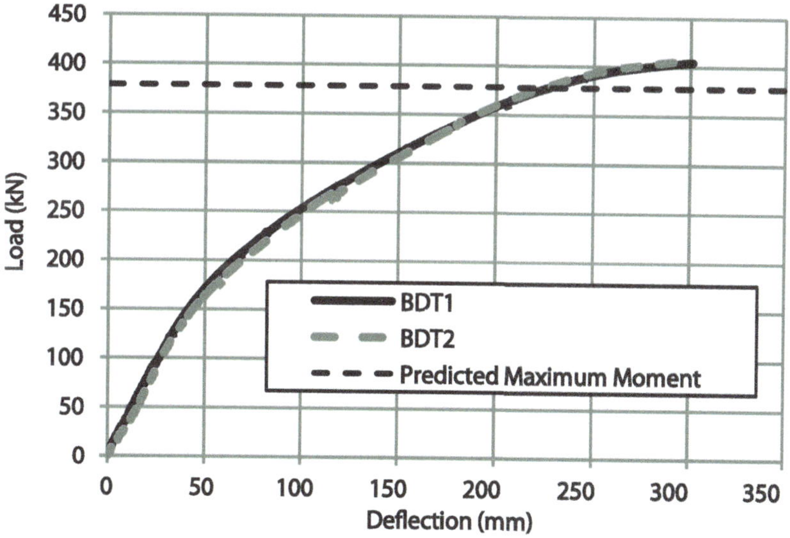

Fig. 11. Load versus deflection plot for midspan flexure tests

5.5 Shear Testing

Figure 12 shows the load versus deflection plots for the shear tests. Strain compatibility was used to predict moment strength, and the 2010 AASHTO LRFD Simplified Design Equation was used for the shear prediction. Each test exhibited similar behavior for the majority of the tests. The difference between the two tests was the addition of a CIP diaphragm surrounding the extended strands to prevent strand slippage. A sudden failure was observed for both specimens; however, the failure modes were distinctly different, as evidenced by the end of the plots in Fig. 12. Each girder held 800 kN plus its dead load, which is equivalent to 2.5 HS-20 design trucks.

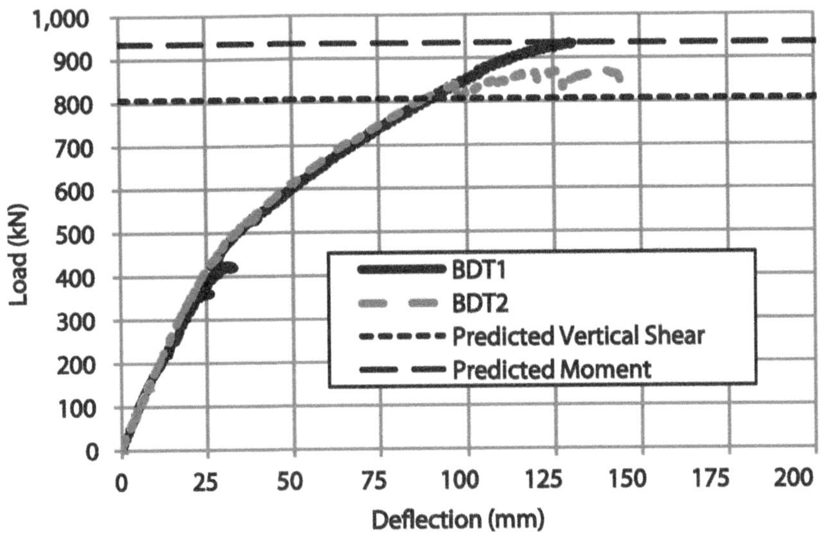

Fig. 12. Load versus deflection plot for shear tests

The BDT1 exhibited a bond failure, shown in Fig. 13, which was determined by the bottom strand slip values, plotted in Fig. 14 (positive slip indicates strand draw-in). The east bottom strand started slipping at around 823 kN. Both gauges slipped well beyond the predetermined 0.25 mm bond failure limit. For this reason, the protruding strands were bent upward and cast into a diaphragm for the second shear test. This addition was expected to force a shear failure.

The sudden failure of BDT2, shown in Fig. 15, was caused by the separation of the CIP topping from the PC section. The failure mode was observed by the ESRG mounted on the CIP concrete interface. The ERSG readings, placed on either side of the CIP/PC interface, are presented in Fig. 16, where tension is positive and compression is negative. Predictably, the strain in both gauges began compressive with the CIP gauge higher than the precast gauge. However, near 800 kN, the PC gauge gained compression without additional load. Conversely, the CIP gauge went into tension in the same manner. This sudden strain divergence indicated composite action up to

Fig. 13. BDT shear test, bond failure

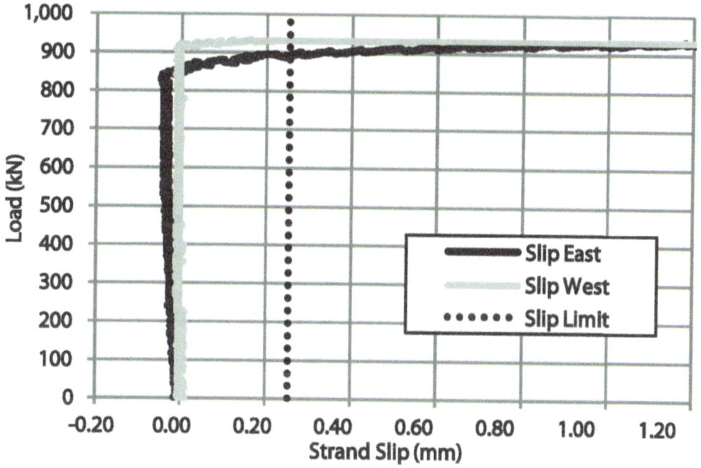

Fig. 14. Load versus slip plot for BDT1 shear test, bottom strands

approximately 800 kips and a sudden shift to non-composite behavior through failure at 814 kN. It was determined that the failure observed for BDT2 was initiated by a horizontal shear failure.

Because the fresh HSSCC is exceptionally difficult to roughen, the hardened surface was sand blasted prior to the deck placement. This technique is common practice for many building products when a CIP topping is used. In this instance of very high horizontal shear stresses, improbable under realistic circumstances, this surface preparation was not enough. Also indicated in Fig. 16 are the AASHTO horizontal shear capacities converted to applied point loads. Capacities were calculated using roughened and smooth concrete interface coefficients and converted to point loads for

Fig. 15. BDT2 horizontal shear failure

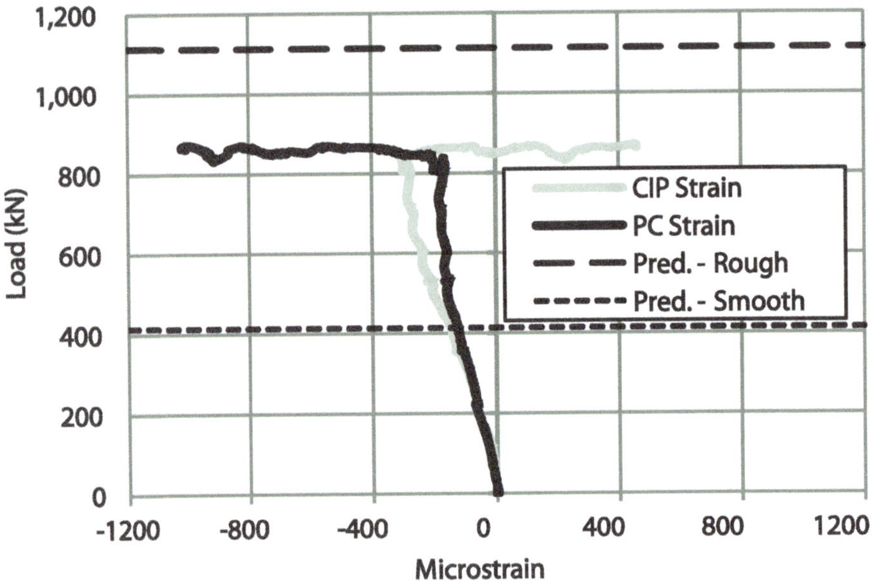

Fig. 16. Load versus longitudinal strain on either side of CIP/PC interface

comparison purposes. Clearly, the girder did not achieve the roughened interface capacity. It is conservatively recommended that the smooth concrete shear friction coefficients be used during design with a sand blasted interface.

5.6 Testing Summary

Table 4 summarizes calculated predictions and test values. Each prediction accounted for dead load, flexure predictions utilized strain compatibility, shear calculations employed the AASHTO Simplified Shear Method and horizontal shear used AASHOT LRFD shear friction predictions. Both girders achieved the predicted moment capacity for the development length tests indicating harped 18 mm diameter strand development at the AASHTO LRFD predicted development length. The flexure tests resulted in slightly higher than the calculated capacities with a test-to-calculated ratio of 1.06. Interestingly, the BDT1 shear test nearly resulted in a fully developed flexural failure with a test-to-calculated moment ratio of 1.0 indicating the 1829 mm strand embedment length may be near the true development length.

Table 4. Flexure and shear test summaries for BDT specimen

		M_{test}, kN-m	M_{calc}, kN-m	V_{test}, kN	V_{calc}, kN	V_{htest}, kN/m	V_{hcalc}, kN/m	Failure type
Flexure test	BDT1	1,750	1,650	245	645	420	938–2600	Flexure
	BDT2	1,750		249		420		Flexure
Shear test	BDT1	1,480	1,480	885	770	2090	938–2600	Bond
	BDT2	1,380		814		1950		Horizontal shear

Because the exceptional compressive and bond strengths of the HSSCC creates cracks that pass directly through the aggregate, past concerns have been raised that this behavior would eliminate or reduce aggregate interlock and friction, which aid vertical shear resistance. These are only two tests, but even with the bond and horizontal shear failures, both girders exceeded the shear predictions. The shear tests resulted in test-to-calculated shear ratios of 1.05–1.15. However, neither tests resulted in a vertical shear failure, suggesting additional shear capacity was available. Previous tests (Morcous et al. 2011b) have also indicated adequate shear prediction and capacity for similar HSSCC with WWR reinforcement.

The horizontal shear failure of BDT2 occurred at a load much higher than probable on the single tee girder. The observed horizontal shear capacity for BDT2 (2056 kN/m) fell between the AASHTO LRFD smooth (934 kN/m) and roughened (2596 kN/m) interface predictions. This indicates that the sand blasted surface preparation was not enough to sustain the intentionally roughened horizontal shear. Design should conservatively use the smooth interface prediction. BDT1 attained a higher load and failed in bond rather than horizontal shear.

6 Summary and Conclusions

The purpose of this research was to develop a simple bridge system combining 18 mm diameter prestressing strands, HSSCC and WWR in standard precast products for use as practical short and medium spans. This girder represents one of the first uses of 18 mm diameter prestressing strands and the first in which these strands were harped.

The standard PCI "heavy" double tee girders were selected because of their cost effectiveness and production efficiency. In addition, 18 mm diameter prestressing strands and WWR were used to significantly decrease labor. Several experimental investigations were carried out on two 15.5 m long and 603 mm deep single tees (two halves of one double tee girder) to evaluate their behavior. Transfer and development length of the 18 mm diameter strands at 51 mm by 51 mm spacing and shear transfer between the HSSCC girder and CIP concrete deck was also investigated. Based on these results, the following conclusions were made:

- Transfer length of harped 18 mm diameter strands with the investigated jacking stress in 83 MPa design release strength concrete at 51 mm by 51 mm spacing was measured up to 470 mm. This is significantly shorter than that predicted by ACI-318 (841 mm) and the AASHTO LRFD Specifications (1067 mm).
- Development length of harped 18 mm diameter strands with the jacked to $0.6f_{pu}$, in 103 MPa design strength concrete at 51 mm by 51 mm spacing is conservatively predicted by the AASHTO LRFD Specifications.
- Flexure and shear predictions for the design of I-girders in the AASHTO LRFD Specifications are applicable to 103 MPa design strength concrete BDT girders with harped 18 mm strands at 51 mm by 51 mm spacing. Actual flexure and shear capacities compared well with the predicted values using measured material properties.
- The contribution of the contact surface between PC girder and CIP deck should be conservatively considered as "smooth concrete" unless appropriate interface roughening is achieved.

References

American Association of State and Highway Transportation Officials (AASHTO): LRFD Bridge Design Specifications. AASHTO, Washington, DC (2008)

American Concrete Institute (ACI) Committee 318: Building Code Requirements for Structural Concrete (ACI 318-08) and Commentary. ACI, Farmington Hills, MI (2008)

Barnes, R.W., Grove, J.W., Burns, N.H.: Experimental assessment of factors affecting transfer length. ACI Struct. J. **100**(6), 740–748 (2003)

Carroll, C., Cousins, T.E., Roberts-Wollman, C.L.: Top strand effect. In: PCI National Bridge Conference, Precast/Prestressed Concrete Institute, Chicago, IL (2008)

Maguire, M.: Impact of 0.7 inch diameter prestressing strands in bridge girders. M.S. thesis, University of Nebraska-Lincoln, Lincoln, NE (2009)

Morcous, G., Tadros, M.K.: Application of ultra-high performance concrete to bridge girders. Report to the Nebraska Department of Roads (2009)

Morcous, G., Hanna, K., Tadros, M.K.: Use of 0.7-in.-diameter strands in pretensioned bridge girders. PCI J. Fall, 65–82 (2011a)

Morcous, G., Maguire, M., Tadros, M.K.: Welded-wire reinforcement versus random steel fibers in precast prestressed concrete bridge girders. PCI J., 113–129 (2011b)

A Machine-Learning Approach for Semantic Matching of Building Codes and Building Information Models (BIMs) for Supporting Automated Code Checking

Ruichuan Zhang(✉) and Nora El-Gohary

Department of Civil and Environmental Engineering, University of Illinois at Urbana-Champaign, Urbana, IL, USA
{rzhang65, gohary}@illinois.edu

Abstract. Various automated code compliance checking (ACC) systems have been developed and used to check the compliance of building information models (BIMs) with building codes, to reduce the time, cost, and errors of the code compliance checking process. All these systems require some form of code-BIM matching – matching of the concept representations in the codes to those in the BIMs – which is a difficult task. Traditionally, semantic matching was conducted in a highly-manual manner. To address this problem, more recently, a limited number of efforts have proposed fully automated semantic matching methods, which mostly rely on matching annotations and/or rules developed by domain experts. Despite their relatively good performance, these methods are by nature difficult to generalize or scale up (e.g., the matching rules need to be updated, modified, or extended when switching from one type of code to another). There is, thus, a need for semantic matching approaches that are more generalizable and scalable. To address this need, this paper proposes a new, machine learning-based approach to automatically match the building-code concepts and relations to their equivalent concepts and relations in the Industry Foundation Classes (IFC). The proposed approach consists of five primary tasks: (1) prepare and process the training and testing data; (2) automatically identify the domain word embeddings by learning from a large corpus of building-code text and generate the final semantic representations by combining the domain and general word embeddings; (3) match the building-code concepts to the IFC elements; (4) match the building-code relations to the IFC relations; and (5) evaluate the performance of the proposed approach using accuracy. The proposed approach was implemented and tested on a number of chapters from the 2009 International Building Code (IBC) and the Champaign 2015 IBC Amendments. The preliminary results show that the proposed approach achieved an accuracy of 77% for matching building-code concepts to IFC elements, and 78% for matching building-code relations to IFC relations, indicating promising semantic matching performance.

H. Rodrigues et al. (Eds.): GeoMEast 2019, SUCI, pp. 64–73, 2020.
https://doi.org/10.1007/978-3-030-34216-6_5

1 Introduction

To reduce the time, cost, and errors of compliance checking, various automated code compliance checking (ACC) systems have been developed and used to check the compliance of building information models (BIMs) with building codes. These systems have used different methods for information representation and code checking, and have achieved different levels of automation and performance. However, all of them require some form of code-BIM matching – matching of the concept representations in the codes to those in the BIMs. A certain level of matching can be conducted by simply matching natural language words and/or searching through the domain ontology (e.g., match "beam" to "IfcBeamTypeEnum – Beam"). However, it is difficult to match the regulatory information in building-code concepts represented by phrases and clauses, and building-code relations represented by verbs and/or adjectives, to Industry Foundation Classes (IFC) concepts (e.g., "return through" and "detoxification compound facilities and spaces" each cannot be directly matched to an IFC concept). Thus, to ensure the performance of the ACC systems, there is a need to develop an information matching approach that is capable to deal with regulatory information carried in natural language with diversified syntactic and semantic patterns.

In many cases, semantic matching was conducted in a highly-manual manner. With the increasing opportunities and needs for automation, more recent ACC efforts have, instead, proposed semi-automated (e.g., using machine learning algorithms to identify candidate matches, and requiring a human user/expert to verify these matches) (Zhang and El-Gohary 2016). Most recently, a limited number of efforts have also proposed fully automated semantic matching methods, which mostly rely on matching annotations and/or rules developed by domain experts (Zhou and El-Gohary 2018). Despite their relatively good performance, these methods are by nature difficult to generalize or scale up – when switching from one type of code to another, or from one chapter to another in the same code, the matching rules might need to be updated, modified, or extended. There is, thus, a need for semantic matching approaches that are more generalizable and scalable.

To address this need, this paper proposes a new, data-driven approach to automatically match the building-code concepts and relations to their equivalent concepts and relations in the IFC. The proposed approach consists of five primary tasks: (1) prepare and process the training and testing data; (2) automatically identify the domain word embeddings of the building-code concepts and relations by learning from a large corpus of building-code text and generate the final semantic representations by combining the domain and general word embeddings; (3) match the building-code concepts to the IFC elements using a similarity-based method; (4) match the building-code relations to the IFC relations using a supervised learning-based method; and (5) evaluate the performance of the proposed approach using accuracy.

2 Background

2.1 Semantic Matching

Semantic matching aims to identify the information that is semantically related (Fernández et al. 2011). Many research efforts have been undertaken to match information from sources such as text and information models other than building information models (BIMs) to the information from BIMs. For example, Cemesova et al. (2015) proposed PassivBIM to integrate the geometric and building fabric information from BIMs with the energy information in building performance simulation (BPS) models. Karan et al. (2015) used a semantic web-based method to identify the common entities among BIMs and geographic information (GIS) systems. Zhang and El-Gohary (2016) proposed a semi-automated learning-based method for matching the regulatory concepts and relations extracted from building codes to their most-related IFC concepts (e.g., equivalent concept, subconcept, superconcept) and relations for supporting ACC. Afsari et al. (2017) proposed ifcJSON representations to map information in the IFC data format to information in the JSON data format for facilitating web-based BIM data exchange. Zhou and El-Gohary (2018) proposed a rule-based method for matching the semantic information elements extracted from energy codes to the IFC concepts and relations for supporting energy code compliance checking.

The majority of semantic matching methods require that the semantic similarities between the concepts and relations are first established. To assess the semantic similarities between the concepts and relations – which are in the form of natural language, those concepts and relations need to be first represented in computer-processible semantic representations. Word embeddings is one of the most widely used semantic representations of natural language data. A word embedding is a vector representation of the word in a specific context (e.g., building code) (Mikolov et al. 2013). Word embeddings have been used for solving numerous text analytics tasks both in the computational linguistic domain [e.g., social media text classification (Xiao et al. 2018), semantic discovery (Yao et al. 2018)] and in the construction domain [e.g., building-code requirement analytics (Zhang and El-Gohary 2019)].

2.2 Industry Foundation Classes

The Industry Foundation Classes (IFC) data format aims to describe, represent, exchange, and share information typically used in the Architecture, Engineering, and Construction (AEC) domain, and is the most commonly used format of building information modeling (buildingSMART 2019). The IFC data format defines an object-based information model consisting of IFC elements and IFC relations. An IFC element is a physically existent component of a project in BIM (buildingSMART 2019). The most important IFC elements include the spatial structure elements (e.g., IfcSpace) and the building elements (e.g., IfcDoor). An IFC relation describes how the IFC elements are related to each other. For example, the "spatial composition" relation describes the case where a spatial structure element decomposes into other IFC elements. However, the IFC concepts do not correspond to the concepts and relations used in the building codes, which creates a major barrier for ACC. It makes the process of matching the building-code concepts and relations to the IFC elements and relations complex and challenging.

3 Proposed Machine-Learning Approach for Semantic Matching of Building Codes and Building Information Models

The proposed approach consists of five main steps, as shown in Fig. 1: (1) data preparation and preprocessing, (2) semantic representation generation, (3) similarity-based matching of building-code concepts to IFC elements, (4) supervised learning-based matching of building-code relations to IFC relations, and (5) method evaluation.

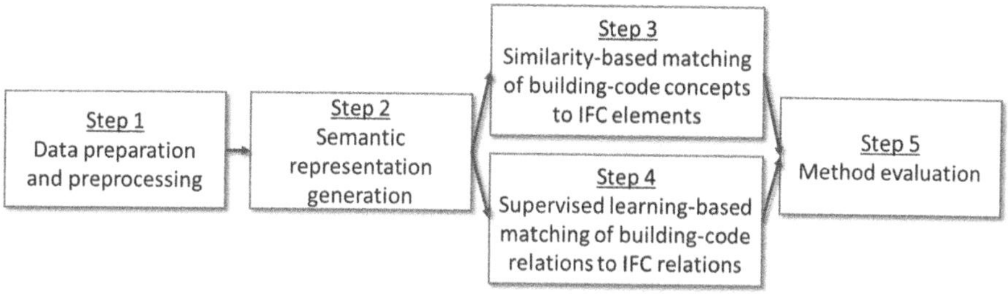

Fig. 1. Proposed machine-learning approach for semantic matching of building-code concepts and relations to Industry Foundation Classes (IFC) elements and relations

Step 1: Data Preparation and Preprocessing

For evaluating the matching of building-code concepts and relations, 80 sentences were selected from the 2009 IBC and the Champaign 2015 IBC Amendments. The concepts and relations in these sentences were manually extracted, resulting in a total of 97 building-code concepts and 73 building-code relations (including the two building-code concepts associated by this building-code relation, Concept A and Concept B). Each word in the names of the concepts and the relations was lowercased and singularized. All the concepts and relations were manually matched to the equivalent or super IFC elements or IFC relations, resulting in matching the 97 concepts to a total of 24 IFC elements and the 73 relations to a total of six IFC relations, as shown in Tables 1 and 2, respectively. The six relations include five original IFC relations, in addition to an added sixth relation, "complex relation", which was added to represent the case where a single building-code relation needs to be matched to multiple IFC relations. Each IFC element was further transformed into a canonical form – which is a lowercased English word, phrase, or sentence – for the purpose of semantic similarity assessment (Step 3). The transformation was conducted following three steps: (1) removing the prefixes in the IFC elements (e.g., "Ifc"), (2) referring to the explanations provided by the IFC documentation, and (3) using engineering judgment. For example, "IfcDoor" was transformed into "door", and "IfcSpace" was transformed into "room, space, or unit".

Table 1. Industry Foundation Classes (IFC) elements used in the proposed semantic matching approach

Type of IFC elements	IFC elements
Spatial structure elements	IfcSite, IfcBuilding, IfcBuildingStorey, IfcSpace
Building elements	IfcBeam, IfcChimney, IfcColumn, IfcCovering, IfcCurtainWall, IfcDoor, IfcFooting, IfcMember, IfcPile, IfcPlate, IfcRailing, IfcRamp, IfcRoof, IfcSlab, IfcStair, IfcWall, IfcWindow
Furnishing elements	IfcFurniture
Transportation elements	IfcTransportElementTypeEnum – Escalator, IfcTransportElementTypeEnum – Elevator

Table 2. Industry Foundation Classes (IFC) relations used in the proposed semantic matching approach

IFC relations	Definitions
Spatial composition	A spatial structure element decomposes into other IFC elements
Spatial container	A spatial structure element contains other IFC elements
Product placement	An IFC element's location relative to another IFC element
Material constituent	An IFC element consists of a material element
Property	An IFC element has a property

Two types of data were prepared for generating the semantic representations (Step 2). For identifying the domain word embeddings, a corpus of 6,000 sentences from the 2009 IBC and the Champaign 2015 IBC Amendments were used to train an unsupervised learning algorithm. For the general word embeddings, the "pre-trained word embeddings" (Pennington et al. 2014) were used. Those word embeddings were learned from a large, cross-domain corpus, using the Glove algorithm, and thus can provide additional semantic information (Pennington et al. 2014) to enhance the robustness of the semantic representations.

Step 2: Semantic Representation Generation

The semantic representations of the building-code concepts and relations, and the IFC elements and relations, were generated based on word embeddings. The semantic representation generation step consists of two substeps: (1) training the learning algorithm for identifying the domain word embeddings, and (2) combining the domain and the general word embeddings. First, the unsupervised learning algorithm, word2vec, was trained on the domain-specific corpus of building-code sentences using the Gensim (Rehurek and Sojka 2010) built in Python, in order to identify the domain word embeddings. Second, for each word, the final word semantic representation was computed as the weighted average of the domain and general embeddings, in order to reflect both the domain-specific semantic meanings and the general semantic meanings of the word. The domain semantic weight ranges from 0 to 1, where 0 represents only using general semantic meanings and 1 represents only using domain semantic meanings.

Step 3: Similarity-Based Matching of Building-Code Concepts to IFC Elements
The building-code concepts were matched to the IFC elements using a similarity-based method, which consists of two substeps: semantic similarity assessment and concept-element matching. First, the semantic similarities between the building-code concepts and the canonical forms corresponding to the IFC elements were computed. Two semantic similarities were proposed: phrase similarity and last-word similarity. Phrase similarity is defined as the cosine similarity between the phrase semantic representations of the building-code concept and the canonical form of the IFC element. A phrase semantic representation is formed by averaging the word semantic representations of all words in the building-code concept or the canonical form. Last-word similarity is defined as the cosine similarity between the semantic representation of the last noun in the building-code concept – which typically carries important information about building elements – and the phrase semantic representation of the canonical form of the IFC element. Second, the building-code concepts were matched to the IFC elements based on matching scores. For each pair of building-code concept and IFC element, the higher one of the phrase and last-word similarities was used as the matching score. For each building-code concept, the candidate IFC element having the highest matching score was selected as the match.

Step 4: Supervised Learning-Based Matching of Building-Code Relations to IFC Relations
The building-code relations were matched to the IFC relations using a supervised learning-based method, which consists of two substeps: semantic feature development and relation classification. First, four semantic features were selected: the phrase semantic representations of the relation, Concept A, and Concept B, and the lettercase of the words in Concept B. Similar to Step 3, the phrase semantic representations were computed as the average of the semantic representations of the words in the relations and the concepts. The fourth feature (i.e., the lettercase of the words in the object) is binary, indicating whether there is a capitalized word in the object. Second, a relation classification model was trained using the training data. Two types of classifiers were tested and compared: a multilayer perceptron (MLP) and a multiclass support vector machine (SVM) with a linear kernel. The trained relation classification model is able to take new features and predict the corresponding IFC relations automatically.

Step 5: Evaluation
The performances of matching building-code concepts to IFC elements and matching building-code relations to IFC relations were evaluated separately, both using accuracy (Olson and Delen 2008). Accuracy is defined as the proportion of the testing building-code concepts or relations that are correctly matched to their corresponding IFC elements or relations, in the entire testing building-code concepts or relations dataset.

4 Preliminary Experimental Results

4.1 Performances in Code-BIM Matching

The performance of the proposed approach is summarized in Table 5. Based on the testing results, the accuracy of matching the building-code concepts to the IFC elements is 77% and the accuracy of matching the building-code relations to the IFC relations is 78%. Examples of the correctly matched pairs of building-code concepts and IFC elements, and pairs of building-code relations (with associated concepts) and IFC relations, are shown in Tables 3 and 4, respectively.

Table 3. Example matched building-code concepts and Industry Foundation Classes (IFC) elements

Building-code concepts (A)	Matched IFC element (B)	Type of match (relation of B to A)
Horizontal sliding power-operated door	IfcDoor	Superconcept
Building	IfcBuilding	Equivalent concept
Permanently installed furnishing	IfcFurniture	Superconcept
Mezzanine	IfcFloor	Superconcept
Type A dwelling unit	IfcSpace	Superconcept

Table 4. Example matched building-code relations and Industry Foundation Classes (IFC) relations

Building-code relations (with associated concepts)	Matched IFC relations	Type of match
Have, dwelling unit, room	Spatial composition	Equivalent relation
With, room, furred ceiling	Spatial container	Equivalent relation
To, egress, exit	Product placement	Equivalent relation
Have, corridor, ceiling height	Property	Equivalent relation
Accessory to, area, area	Complex relation	Equivalent relation

For matching the building-code concepts to the IFC elements and matching the building-code relations to the IFC relations, different domain semantic weights were tested and compared, including 0 (using general word embeddings only), 0.25, 0.50, 0.75, and 1 (using domain word embeddings only), as shown in Table 5. The optimal performance for matching the building-code concepts to the IFC elements was achieved when the domain semantic weight was 0.25; and the optimal performance for matching the building-code relations to the IFC relations was achieved when the domain semantic weight was 0.50. Compared to using only either domain word embeddings or general word embeddings, the use of weighted word embeddings (i.e., the proposed semantic representation) increased the accuracies by up to 22%, which indicates the benefit of integrating both domain-specific and cross-domain semantic information.

For matching the building-code relations to the IFC relations, the two tested classification algorithms (i.e., MLP and SVM) achieved different performances for different semantic weights, but achieved the same optimal performance when the domain semantic weight is 0.50, as shown in Table 5.

4.2 Error Analysis

Two main types of errors were identified based on the experimental results. First, for matching building-code concepts to IFC elements, the proposed method had errors when dealing with building-code concepts that are less frequently appearing in the building code, such as “casework”, which appears only once in the entire IBC 2009. The generated domain word embeddings may not be able to capture the domain semantic meanings of those concepts. In future work, a larger, more diversified corpus of text from the construction domain could be used for training. Second, for matching the building-code relations to the IFC relations, the proposed method misclassified “spatial composition” as “spatial container” or “complex relation”. In future work, more training data and features based on domain ontology could be used, in order to enhance the ability of the relation classification model to distinguish such relation types that are close or related.

Table 5. Performance of the proposed approach with different domain semantic weights

Domain semantic weights	Accuracy of matching building-code concepts and Industry Foundation Classes (IFC) elements[a]	Accuracy of matching building-code relations and IFC Relations[a]	
		Multilayer perceptron	Multiclass support vector machine
0	76%	61%	56%
0.25	**77%**	69%	74%
0.50	71%	**78%**	**78%**
0.75	70%	69%	69%
1	66%	65%	69%

[a]Bolded font indicates the highest performance

5 Conclusions

This paper proposed a new machine learning-based approach for matching semantic information in building codes and building information models for supporting automated compliance checking (ACC), by separately matching the building-code concepts to the IFC elements and matching the building-code relations to the IFC relations. First, the semantic representations were generated by combining the domain word embeddings and the general word embeddings to reflect both domain-specific and cross-domain semantic information, in order to improve both accuracy and scalability of the proposed approach. Second, a similarity-based method was proposed to match the building-code concepts to the IFC elements. Third, a supervised learning-based method was proposed to match the building-code relations to the IFC relations. The proposed approach achieved a 76% accuracy of matching the building-code concepts to the IFC elements, and a 78% accuracy of matching the building-code relations to the IFC relations.

This paper contributes to the body of knowledge in two primary ways. First, the paper proposed a new way to model the semantic meanings of the domain-specific text by first generating the domain word embeddings and then combining both the domain and the general word embeddings. The proposed approach makes use of both domain and general semantic representations in semantic matching, and thus has potentially better scalability in dealing with different types of building codes. Second, the initial experimental results show that the proposed semantic representation successfully captured the semantic meanings of both building-code concepts and relations, and IFC elements and relations, in both similarity-based and supervised learning-based semantic matching tasks.

In their future work, the authors first plan to improve the information matching by including more IFC elements (e.g., IfcSanitaryTerminal), the properties of the IFC elements (e.g., Pset_DoorCommon – IsExternal), the subconcepts of the IFC elements [e.g., revolving door (a subconcept of IfcDoor)], and more IFC relations (e.g., element filling); and including building-code concepts and relations described in complex phrases (e.g., occupant evacuation elevator lobby) and sentences. Second, the authors will explore further ways to improve the performance of the proposed information matching approach, including using more training data for the domain word-embedding generation, annotating more training data for relation classification, and exploring different data similarities for matching the building-code concepts to the IFC elements, and different supervised learning algorithms for matching the building-code relations to the IFC relations. Third, and most importantly, the authors plan to integrate the proposed information matching approach with machine learning-based information extraction and transformation approaches, with an aim to develop a fully automated, and highly scalable ACC system.

Acknowledgments. The authors would like to thank the National Science Foundation (NSF). This material is based on work supported by the NSF under Grant No. 1827733. Any opinions, findings, and conclusions or recommendations expressed in this material are those of the authors and do not necessarily reflect the views of the NSF.

References

Zhang, J., El-Gohary, N.: Extending building information models semiautomatically using semantic natural language processing techniques. J. Comput. Civ. Eng. ASCE (2016). https://doi.org/10.1061/(asce)cp.1943-5487.0000536

Zhou, P., El-Gohary, N.: Automated matching of design information in BIM to regulatory information in energy codes. In: Construction Research Congress 2018: Construction Information Technology, ASCE (2018)

Cemesova, A., Hopfe, C.J., Mcleod, R.S.: PassivBIM: enhancing interoperability between BIM and low energy design software. Autom. Construct. (2015). https://doi.org/10.1016/j.autcon.2015.04.014

Karan, E.P., Irizarry, J., Haymaker, J.: BIM and GIS integration and interoperability based on semantic web technology. J. Comput. Civ. Eng. ASCE (2015). https://doi.org/10.1061/(asce)cp.1943-5487.0000519

Afsari, K., Eastman, C.M., Castro-Lacouture, D.: JavaScript Object Notation (JSON) data serialization for IFC schema in web-based BIM data exchange. Autom. Construct. (2017). https://doi.org/10.1016/j.autcon.2017.01.011

Fernández, M., Cantador, I., López, V., Vallet, D., Castells, P., Motta, E.: Semantically enhanced information retrieval: an ontology-based approach. J. Web Semantics (2011). https://doi.org/10.1016/j.websem.2010.11.003. Special Issue on Semantic Search

Yang, X., Macdonald, C., Ounis, I.: Using word embeddings in twitter election classification. Inf. Retrieval J. (2018). https://doi.org/10.1007/s10791-017-9319-5

Yao, Z., Sun, Y., Ding, W., Rao, N., Xiong, H.: Dynamic word embeddings for evolving semantic discovery. In: Proceedings of the Eleventh ACM International Conference on Web Search and Data Mining, ACM (2018). https://doi.org/10.1145/3159652.3159703

Zhang, R., El-Gohary, N.: A machine learning-based approach for building code requirement hierarchy extraction. In: Proceedings of the 7th CSCE International Construction Specialty Conference (Jointly with Construction Research Congress), CSCE (2019)

Pennington, J., Socher, R., Manning, C.D.: GloVe: global vectors for word representation. In: Proceedings of the 2014 Conference on Empirical Methods in Natural Language Processing, ACL (2014). https://doi.org/10.3115/v1/d14-1162

Rehurek, R., Sojka, P.: Software framework for topic modelling with large corpora. In: LREC 2010 Workshop on New Challenges for NLP Frameworks, ELRA (2010)

Olson, D.L., Delen, D.: Advanced Data Mining Techniques. Springer, Heidelberg (2008). https://doi.org/10.1007/978-3-540-76917-0

buildingSMART: Industry Foundation Classes, Version 4 - Addendum 2, 15 June 2019. http://www.buildingsmart-tech.org/ifc/IFC4/Add2/html/

EICP Cemented Sand Modified with Biopolymer

Mohamed Arab[1,2], Maher Omar[1], Roudha Aljassmi[1], Reem Nasef[1(✉)], Lubna Nassar[1], and Sham Miro[1]

[1] Department of Civil and Environmental Engineering, College of Engineering, University of Sharjah, 27272, Sharjah, UAE
marab@sharjah.ae.ac, {momar,U15102109}@sharjah.ac.ae
[2] Structural Engineering Department, Mansoura University Al-Gomhoria Street, Mansoura, Egypt
mg_arab@mans.edu.eg

Abstract. Urbanization growth in the world in past two decades are considered a challenge in terms of supplying required construction building materials needed for this development. Bricks are an important construction material that consumes energy, aggregates and cement during their manufacturing. Cement, widely used binder to manufacture brick, is responsible for producing around 5% of manmade CO2 emissions. This research investigates development of a natural bio-cemented sandstone that can be used as a green-bricks. In this approach, urease enzyme will be used to catalyze hydrolysis of urea in presence of calcium to introduce carbonate precipitation that binds Silica sand particles through a process called Enzyme Induced Carbonate Precipitation (EICP). Sodium alginate (SA) biopolymer was used to enhance the performance of the EICP cementation process by promoting carbonate precipitation around sand particles. Different percentages of SA were used to optimize the cementing solution concentrations and come up with a preferred solution for brick manufacturing. Four-point flexural tests and unconfined compression test are used to demonstrate the performance of the proposed approach and to assess its feasibility after curing time of 7 days. Results show these proposed bio-blocks can have strengths ranging from 1 MPa to 2 MPa. Bio-blocks are comparable in terms of stress and stiffness to bricks prepared with cement conventionally used in the block construction.

1 Introduction

Population explosion and migration of people to urban area has created a huge gap between demand and supply of construction materials (viz., brick, cement etc.), leading to abuse of natural resources and increase in CO2 emissions. Generally, bricks are a major construction element that constitutes a high percentage of any structural development. Conventional brick production consumes a large amount of energy and considered one of the main contributors to greenhouse gases emissions Murmu and Patel (2018). Among other alternatives, calcite precipitation has been intensively investigated as a sustainable alternative. Montoya and Dejong (2015)—explains that there are several techniques used to induce calcite precipitation, including urea hydrolysis,

H. Rodrigues et al. (Eds.): GeoMEast 2019, SUCI, pp. 74–85, 2020.
https://doi.org/10.1007/978-3-030-34216-6_6

microbial denitrification and oxidation of organic matter under aerobic conditions or through denitrification.

The hydrolysis of urea is the most advanced mechanism to induce calcite precipitation in terms of development, and most often discussed in the literature due to simplicity Khodadadi et al. (2017). Urea hydrolysis is a reaction in which urea reacts with water to form ammonium and carbonate ($CaCO_3$) irreversibly. Research by Benini et al. (1999) concluded that the Hydrolysis of urea in presence of catalyst (urease Enzyme) is 1014 times faster than chemical (un-catalyzed) urea hydrolysis. Calcium carbonate precipitation improves the strength, stiffness, and dilatancy of the soil by pore filling, molecule roughening, and interparticle binding. In this process, the formation of $CaCO_3$ mimics the bioformation in nature, in contrast to typical chemical techniques. Among other techniques Enzyme Induced Carbonate Precipitation (EICP) is a bio geotechnical technique in which calcium carbonate is precipitated from solution between soil particles for improving the engineering properties of granular soil. Free urease enzyme derived from agricultural source catalyzes the hydrolysis of urea in the solution, which results in carbonate ion production. In the presence of calcium ions from any salt source, the carbonate ions precipitate as calcium carbonate. As shown in Fig. 1 the main purpose is to convert the cohesion-less sand into sandstone.

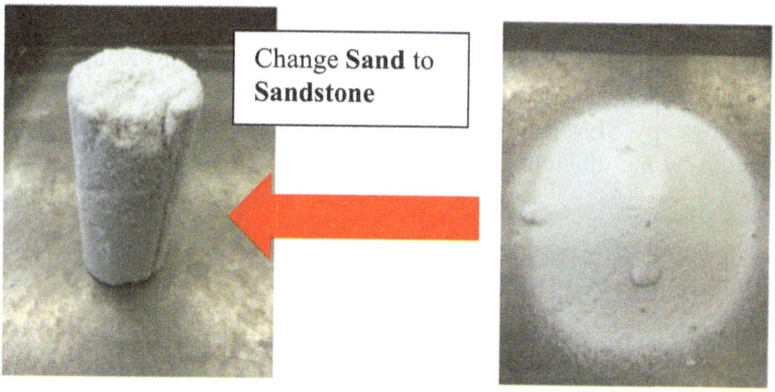

Fig. 1. EICP technique to convert sand to sandstone

In EICP, free urease enzyme catalyzes the hydrolysis of urea in an aqueous solution, forming carbonate ions and alkalinity that leads to calcium carbonate precipitation as shown in the schematic diagram in Fig. 2.

Almajed et al. (2017) have tested several treatment concentrations for the EICP to find the most suitable approach and mix in improving the treated sand uniaxial compressive strength (UCS). The outcome was that an EICP solution concentration of between 0.67 M calcium chloride with 1 M of urea and 3 g/l urease enzyme was sufficient to improve carbonate precipitation and enhance the strength of the soil. More recently, a study showed that introducing non-fat powdered milk as stabilizer was added into the EICP treatment provided nucleation points for carbonate precipitation Almajed et al. (2019). Comparing the UCS of the specimens with powdered milk added

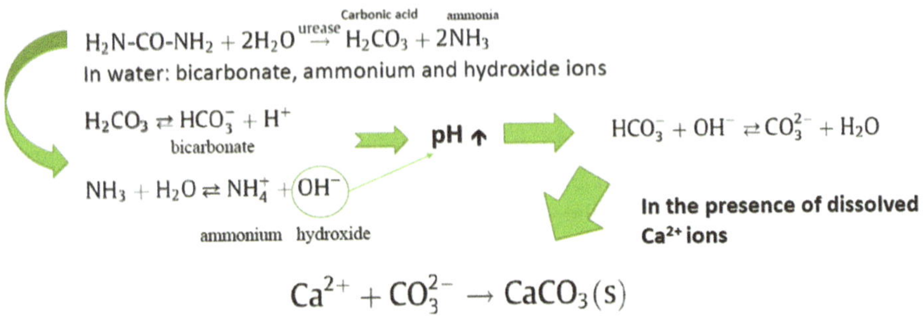

Fig. 2. Summary of the biochemical reaction for the carbonate precipitation in EICP.

to the specimens without powdered milk, resulted in UCS between 1.65 MPa to 1.82 MPa and 0.12 MPa and 0.16 MPa respectively at similar carbonate content Almajed et al. (2018).

In this study, a series of small-scale laboratory experiments were conducted to investigate the feasibility of using a mix of EICP and sodium alginate solutions in soil treatment for purposes of create bio-bricks. The treatment was achieved by mixing EICP, biopolymer mix with the soil to enhance the mechanical properties of the soil. The feasibility was evaluated by different means both mechanical and micro-level physical testing. The UCS of treated soil samples along with flexural strength were used to evaluate the mechanical performance of EICP, and SA treated sands. Scanning electron microscopy SEM analysis techniques was also used to further analyze the microstructure by the different treatments, and the mechanism of treated sands at the micro-level.

2 Materials and Methodology

2.1 Sandy Soil

Graded silica sand ASTM C778 ($e_{max} = 0.95$, $e_{min} = 0.66$, $D_{50} = 0.36$, $D_{10} = 0.2$, $G_s = 2.65$) was used throughout this study.

2.2 EICP Cementing Solution

The EICP solution is comprised mainly of urea [$CO(NH_2)_2$], calcium chloride ($CaCl_2$), urease enzyme, and powder milk as stabilizer agent. The free urease enzyme which is used as a catalyst for the ureolysis process was purchased from Fisher Chemical (derived from jack beans) and had low activity (1500 U/g) according to the manufacturer.

The EICP solution was prepared using deionized water in two separate halves; the urea-calcium chloride solution and the urease enzyme solution to prevent the hydrolysis reaction to commence before treatment (Table 1).

SA is a linear natural polysaccharide derivative of alginic acid comprised of 1,4-β-d-mannuronic (M) and α-l-guluronic (G) acids is widely distributed in organisms as

Table 1. EICP cementing solution concentration levels.

Name of cementing solution	Urea (M)	Calcium chloride (M)	Milk (g/L)	Enzyme (g/L)
2M	2	1.5	8	6

diverse marine brown algae and bacteria Kulkarni et al. (2010). Presence of divalent ions (especially calcium) promote gel formation with exchange of SA monovalent ions, the reaction proceeds almost immediately, changing from a low viscosity solution to a gel structure.

2.3 Preparation of the Samples for UCS

Clear acrylic tubes that are 100 mm long and 50 mm inner diameter was used to prepare soil specimens for UCS testing. The columns were sealed with plastic caps from one side. The soil specimens (353.5 g of sand) were first mixed with the SA with different percentages (dry mix) varying (0.5% and 1.0%). The amount of dry sand corresponding dry density of about 1600 kg/m^3, was quickly mixed with around 70 ml of EICP solution, which is considered about one pore size of sand specimen. The sand/SA/EICP mixture was then placed inside the acrylic tube in three lifts with compaction using a steel rod to reach the target density, after compaction of each lift the surface was scratched prior adding the next left. Duplicates were prepared for each test conducted throughout the study. After compaction, the treatment solution was always a few millimeters above the soil surface, indicating that the packed soil was in a near saturated condition. The top of each column was covered during curing in order to limit the solution evaporation. The samples were left to cure for 4 days with the cementing solution (inside the acrylic tubes). After curing, the EICP solution was drained by gravity from the bottom by creating a hole in the base of the tube and the samples were washed with one pore volume of fresh (deionized water) water to get rid of any apparent cementation by salts. After washing, the specimen was carefully extracted from the tubes. The extracted samples were left to dry inside the oven, at 35 ° C for 24 h. Figure 3 shows the samples after preparation.

2.4 Preparation of the Flexural Test Specimens

The same procedure for preparation the samples for the UCS test was followed. However, for the flexural test the samples were compacted in 3D printed molds that were prepared and designed for this purpose. The inner dimensions of the mold were 200 × 50 × 50 mm. Figure 4 shows the picture of the 3D printed molds. Around 900 g of sand was mixed with 160.5 ml of EICP solution.

2.5 Unconfined Compressive Strength (UCS)

The UCS tests were performed in accordance with ASTM D 2166 (ASTM 2003), applied at a rate of 0.50% axial strain/minute.

Fig. 3. Prepared acrylic tubes for UCS test

Fig. 4. 3D printed molds prepared for the flexural testing

2.6 Three-Point Flexural Tests

The three-point bending tests were conducted to study the flexure behavior of EICP-treated bio-specimen. The specimens were located on two adjusted supports that were 152.4 mm apart from each other, and the vertical load was applied on top one support at middle of the specimen with 50.8 mm distance. Following testing method of ASTM D6272 (ASTM 2010), the vertical load was conducted under strain-controlled conditions at a uniform loading rate of 1.5%/min until beam fails. Figure 5 shows the setup used for the flexural test conducted.

2.7 Micro-scale Testing

SEM analysis was performed in order to have a visual evidence of the inter-particle cementation in general and detect the presence of carbonate precipitation in case of EICP/SA treatment. The location of precipitation is an essential parameter in determining the efficiency of the treatment. SEM images were taken using the Field

Fig. 5. Treated soil specimen during flexural test

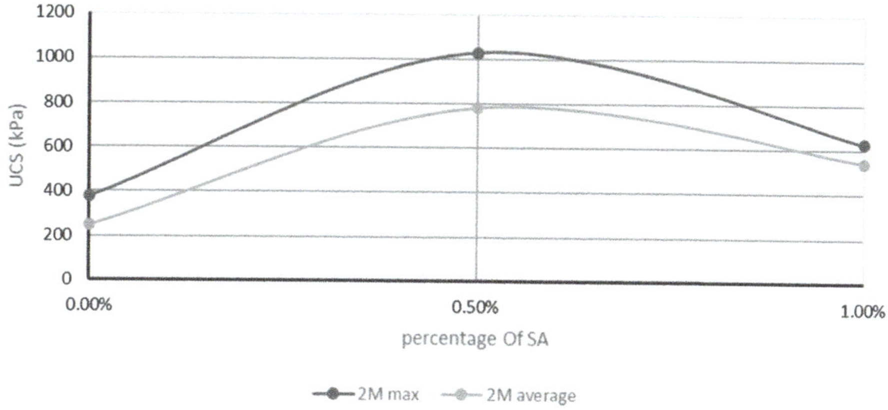

Fig. 6. UCS test results of Sand soil treated with 2M EICP solution vs. SA content

Emission Scanning Electron Microscope (FE-SEM). The specimens were coated with a thin layer of gold for charge dissipation.

3 Results

3.1 UCS Results

Summary of the UCS results of soil specimens treated using the three different solutions are presented in Table 2. The result shows a clear increase in compressive strength with the increase of SA content. However, Fig. 8 shows an optimum SA content at which the compressive strength of the treated soil reaches its maximum value. In case of 2M EICP solution the compressive strength reaches Maximum of

Fig. 7. Tensile splitting of EICP cylindrical specimens treated with mix of EICP solution and SA

Table 2. UCS peak results for treated sand.

UCS (kPa)				
EICP	SA Content	0%	0.5%	1%
2M	Sample 1	117.14	534.76	626.43
	Sample 2	376.88	1028.78	458.37
	Average	247.00	781.77	542.40
	Maximum	376.88	1028.78	626.43

1.1 MPa at SA content of 0.5%. By increasing the SA content to only 0.5% the compressive strength increases by 270%.

Observations of the failure pattern of the specimens following UCS loading indicated that the specimens treated using solutions with SA failed by tensile splitting, as shown in Fig. 7, whereas the specimens without SA failed in shear (Fig. 6).

3.2 Flexure Test Results

Table 3 shows a summary of the results of the flexural test conducted on the beam elements. Comparing the flexural strength of the specimens treated using the EICP solution with SA to the specimens treated without SA, resulted in an average flexural strength between 431 kPa to 720 kPa and 488 kPa respectively. As shown in Table 3, the optimum percentage of SA was 1% of for specimens treated with 2M EICP cementing solution, which resulted in an average value of 510 kPa. Figure 8 shows the increase of the peak flexural stress with the increase of SA content until a maximum increase reached at SA content of 1%.

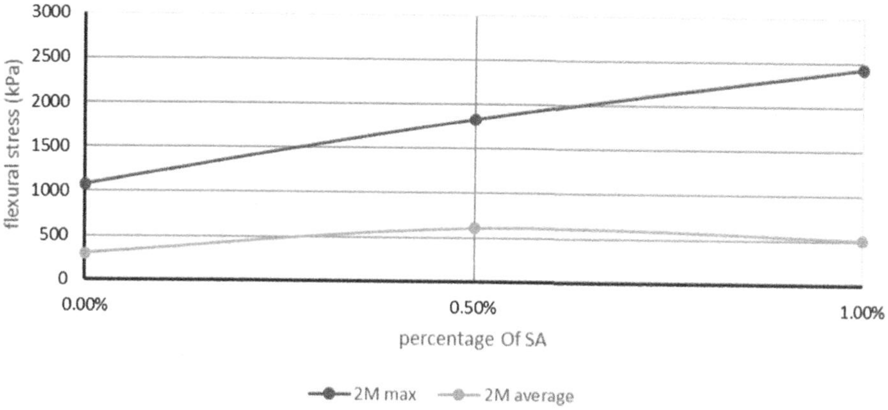

Fig. 8. FS test results of Sand soil treated with 2M EICP solution vs. SA content.

Table 3. Flexural test peak results.

Flexure stress (kPa)				
	Percentage	0%	0.5%	1%
2M	1	1075	1825	2431
	2	980	1342	2200
	Average	300	610	510
	Maximum	1075	1825	2431

4 Discussion

4.1 Comparison with Cement Brick

Figure 9 shows comparison between the flexural stress achieved in the current study using EICP and SA mix for specimens treated with 2M solution with 1% of SA and cement bricks prepared with 10% cement content and water to cement ratio (w/c) = 0.45. The flexural stress results achieved with 2M EICP solution at 1% SA were nearly double the flexural stress achieved from the cement bricks. The soil treated with 2M EICP solution achieved a flexural stress of 2431 kPa. The 10% cement brick achieved a flexural stress of 1300 kPa.

4.2 Comparison with Bu et al. (2018) Results

Bu et al. (2018) investigated the feasibility of using Microbial induced carbonate precipitation (MICP)-treated to build brick. Bu et al. (2018) compared the flexural strength of MICP-cemented samples with cement and lime treated specimens. Comparison was conducted between the results achieved in the current study and the compressive strength for treated sand reported by Bu et al. (2018). Figure 10 shows comparison between the flexural stress achieved in the current study using EICP and

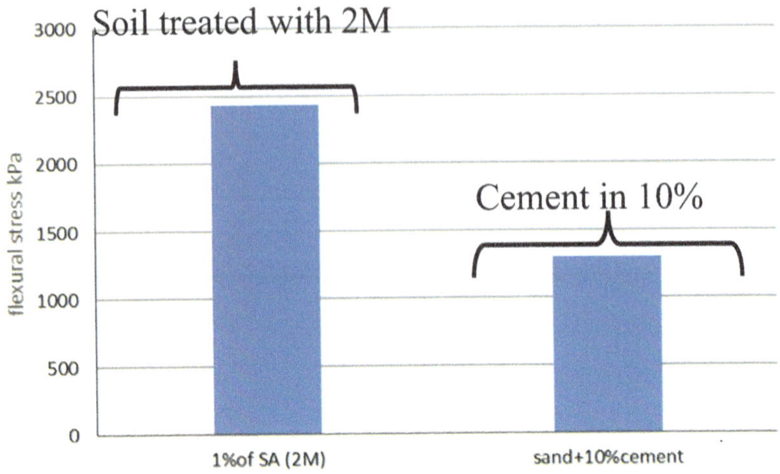

Fig. 9. Comparison between the flexural stresses achieved in the current study by the specimens treated with 2M solution with 1% of SA and cement bricks.

SA mix and sand treated with 20% cement, 20% lime, 30% lime and sand treated with MICP. The results from the current study shows that flexural stress achieved with 1% SA and 2M EICP solution was higher than values reported by the previous studies.

Figure 11 shows comparison between the UCS achieved in the current study using EICP and SA mix and sand treated with20% lime and sand treated with MICP. The results from the current study shows that UCS achieved with 0.5% SA and 2M EICP solution was significantly higher than values reported by Bu et al. (2018).

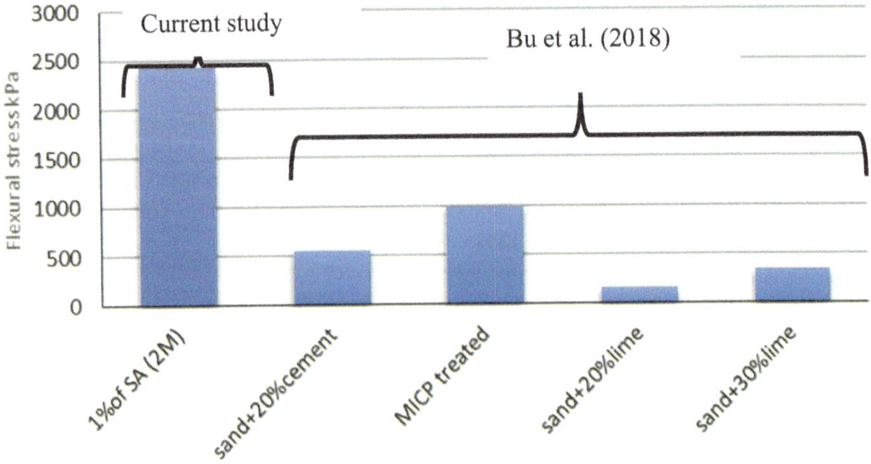

Fig. 10. Comparison between the flexural stress achieved in the current study and values reported by Bu et al. (2018)

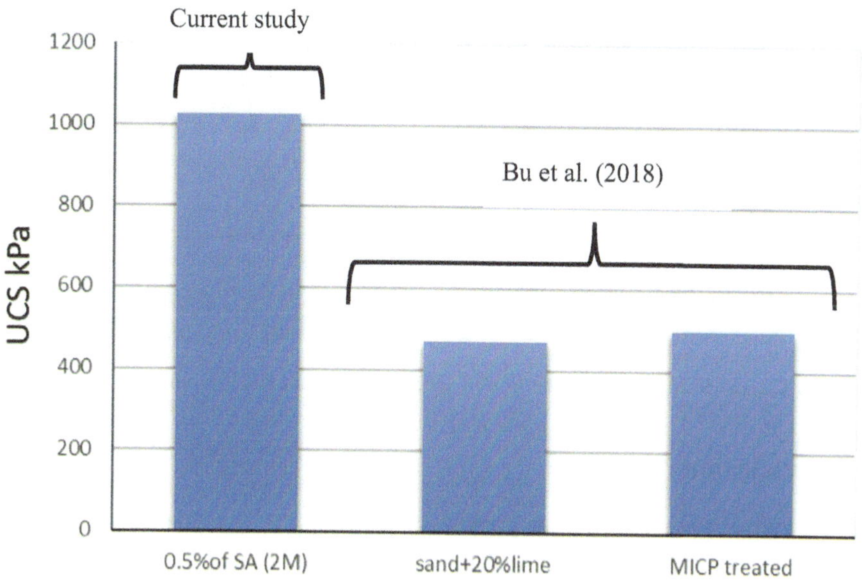

Fig. 11. Comparison between the UCS achieved in the current study and values reported by Bu et al. (2018)

4.3 Micro-level Analysis

SEM analysis was carried out on two specimens treated with 2M EICP solution with 0% and 1% SA. The specimen with 1% SA was chosen for comparison with the soil treated with EICP solution without the SA. Figures 12(a) shows the SEM micrographs of the biopolymer brick. For the specimens treated using the solution that did not contain sodium alginate, the precipitated carbonate appears to be in the form of relatively small crystals distributed over the surface of the sand particle. When SA was added into EICP solution, it appears that relatively large calcite crystals formed, and that the precipitation was focused mainly at inter-particle contacts. The pattern of precipitation is believed to be a major contributor to the increase strength of the specimens prepared with an EICP solution containing SA compared to specimens treated with an EICP solution that did not contain the polymer. The results of SEM testing confirmed precipitation of calcium carbonate in the calcite phase.

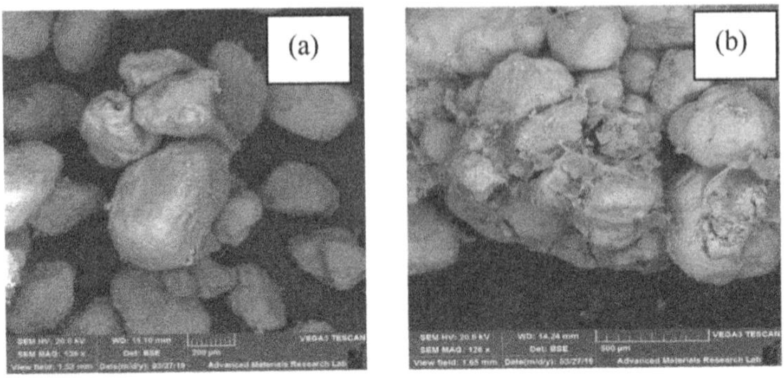

Fig. 12. *(a)* SEM Picture of the treated specimen with 2M EICP (without the SA), *(b)*

5 Conclusion

The feasibility of using SA with EICP to produce biopolymer bricks was studied by conducting unconfined compression tests, flexural tests and SEM imaging analysis. The study investigated the effect of adding sodium alginate on the physical and mechanical properties, composition, and microstructure of the produced biopolymer brick Specimens. The test results presented in this paper show that the addition of sodium alginate to an EICP treatment solution resulted in high unconfined compressive of 1029 kPa for sand treated with 2M EICP solution and 0.5% SA. For flexural strength maximum of 2431 kPa was achieved for sandy soil treated with 2M cementing solution with 1% SA content. SEM imaging was conducted and on the basis of the SEM images, it appears that the mode of soil improvement in these specimens was through inter-particle cementation via carbonate precipitation with the contribution of SA.

References

Almajed, et al.: Baseline investigation on enzyme-induced calcium carbonate precipitation. Geotech. Geoenviron. Eng. J. **144**(11), 04018081 (2018)

Andrew, R.M.: Global CO 2 emissions from cement production. Earth Syst. Sci. Data **10**, 195 (2017). https://doi.org/10.5281/zenodo.831455

Arab, M.G., et al.: Resilient behavior of sodium alginate-treated cohesive soils for pavement applications. Mater. Civ. Eng. J. **31**(1), 04018361 (2018)

Benini, S., et al.: A new proposal for urease mechanism based on the crystal structures of the native and inhibited enzyme from Bacillus pasteurii: why urea hydrolysis costs two nickels. Structure **7**(2), 205–216 (1999)

Hamdan, N., et al.: Hydrogel-assisted enzyme-induced carbonate mineral precipitation. Mater. Civ. Eng. J. **28**(10), 04016089 (2016). https://doi.org/10.1061/(asce)mt.1943-5533.0001604

Khodadadi, T.H., et.al.: Bio-grout materials: a review. In: Grouting 2017 (2017). https://doi.org/10.1061/9780784480793.001

Kulkarni, R.V., et al.: Interpenetrating network hydrogel membranes of sodium alginate and poly (vinyl alcohol) for controlled release of prazosin hydrochloride through skin. Int. J. Biol. Macromol. **47**(4), 520–527 (2010). https://doi.org/10.1016/j.ijbiomac.2010.07.009

Montoya, B.M., Dejong, J.T.: Stress-strain behavior of sands cemented by microbially induced calcite precipitation. J. Geotech. Geoenviron. Eng. **141**(6), 04015019 (2015). https://doi.org/10.1061/(asce)gt.1943-5606.0001302

Murmu, A.L., Patel, A.: Towards sustainable bricks production: an overview. Constr. Build. Mater. **165**, 112–125 (2018). https://doi.org/10.1016/j.conbuildmat.2018.01.038

Zhao, Z., et al.: Biomimetic hydrogel composites for soil stabilization and contaminant mitigation. Environ. Sci. Technol. **50**(22), 12401–12410 (2016)

ASTM, D2166. 2166 (2000): Standard Test Method for Unconfined Compressive Strength of Cohesive Soil. Annual Book of ASTM Standards, American Society for Testing and Materials, Philadelphia 4(08) (2003)

Standard, ASTM D6272–10: Standard Test Method for Flexural Properties of Unreinforced and Reinforced Plastics and Electrical Insulating Materials by Four Point Bending. ASTM International (2010)

Finite Element Analysis Alteration Between the Effect of Seismic and Diverse Wind Speeds on Three Different Resisting Systems

Aya Diab[1(✉)] and Youmn Al-Rawi[2(✉)]

[1] Civil Engineering Department, City University, Tripoli, Lebanon
aya.diab_2@hotmail.com, aya.diab@cityu.edu.lb
[2] Civil Engineering Department, Beirut Arab University, Tripoli, Lebanon
youmn.alrawi@gmail.com

Abstract. The two major loads considered in high-rise buildings are seismic and wind loads. In such buildings, safety means the stability of a structure against the effect of both loads as well as gravity loads. This paper studies the effect of different levels of lateral wind loads on three different seismically designed high-rise buildings on ETABS software. A 40-story high-rise building was seismically designed according to three different resisting systems (special moment resisting frame, shear wall system and dual system). These systems were designed according to a moderate seismic of acceleration 0.25 g and a soil type of Sc which symbolizes the description of a very dense soil and soft rock. Each of these systems were subjected to three global wind speed levels under wind exposure D, low wind level of 45 mph, a moderate level of 85 mph and a high level of 125 mph. The results showed that each system reacts differently not only under the effect of seismic but also under the effect of wind. They also showed that shear wall system failed under the effect of high wind, on the other side it is the most effective system under the effect of seismic and the cheapest one. The dual system is the most desirable system to be used against wind load and the most economical one. Special moment resisting frame system it is desirable to resist high wind load and seismic load but in both cases it costs slightly more than the dual system.

Keywords: Seismic load · Wind load · Shear wall system · Special moment resisting frame · Dual system · High-rise building

1 Introduction

Due to the lack of horizontal spaces nowadays and due to the increase in population, the convenient solution for such a problem is the construction of high-rise building. This increase in the height requires the study of two major loads considered in such tall structures which are the seismic load and the wind load. It is better to know which of these forces is the major one in influencing the structure.

An earthquake is the vibration of earth produced by the rapid release of accumulated energy in elastically strained rocks. The energy released radiates in all directions from its source, the focus. Then this energy propagates in the form of seismic waves.

H. Rodrigues et al. (Eds.): GeoMEast 2019, SUCI, pp. 86–97, 2020.
https://doi.org/10.1007/978-3-030-34216-6_7

To resist this vibration, three systems may be used, which are: the shear wall system, the special moment resisting frame system and the dual system [1].

Wind load is the basic load considered in tall buildings. The knowledge of the dynamic motion it produces and how to resist it is a must for every civil engineer. It depends mostly on the exposed area of the structure that can affect the force exerted on a structure.

2 Literature Review

There are different articles that encounter the subject of this paper and here are some of them:

Application examples for buildings with different heights, floor weights and boundary conditions for both winds and earthquakes such as the intensity of the wind pressure, the seismic zone coefficient, the importance factor, structural system factor and the soil coefficient are analysed and discussed for the purpose of comparison [2].

- Wind is more effective than earthquake for tall buildings with shear walls when minimum design factors are considered, while earthquake was found to be more effective than wind when maximum design factors are considered.
- Earthquake is found to be more effective for short buildings. The wind and earthquake effects increase rapidly when the height of the building increases.

One of them presents a comparative study of the severity effect of earthquake and wind forces on a multi-story building. It is very essential to consider the effects of lateral loads in the design of reinforced concrete structures. It determines the critical design loading for a multi-story building subjected to different basic wind speeds (39, 44, 47, 50, 55 m/s) and earthquake zones (II, III, IV, V). In the present study the response of low-rise and high-rise multi-story buildings are analysed with earthquake loads based on IS 1893:2002 and for wind loads IS 875:1987 code. When a building is situated in a particular wind and earthquake zone, the major designing loads for the particular building can be decided using these results. In this study, the multi-storied building excited to earthquake forces and wind forces for different seismic zones and wind forces are studied. It is concluded that When the lateral forces exerted on high rise buildings, the observed order of the effects are wind speed 55, zone5, wind speed 50, wind speed 47, wind speed 44, zone4, wind speed 39, zone3, zone2 [3].

Here comes an article that studied a high-rise building constructed against seismic in a dual system form. This study evaluates the seismic performance of high-rise buildings primarily designed based on different levels of lateral wind loads (low, moderate and high wind) according to ASCE code. The results showed that the level of design wind load can alter the seismic performance of high-rise dual system buildings. Therefore, even for the cases where the wind demands control the design of lateral load-resisting system, the detailed performance-based seismic evaluation should be carried out to ensure the overall structural safety and integrity [4].

Finally, the paper presented studied the behaviour of two different heights of high-rise building (54 stories, 76 stories) under multi-hazard loadings. The fundamental differences between wind and earthquake demands were highlighted. The results indicated that earthquake loads excite higher modes that produce lower inter-story drift, compared to wind loads, but higher accelerations that occur for a shorter time. Wind-induced accelerations may have comfort and serviceability concerns, while excessive inter-story drifts can cause safety issues. It also shows that high-rise buildings designed for wind may be safe under moderate earthquake loads, regarding the main force resisting system [5].

A 40-story reinforced concrete (RC) building designed on ETABS to resist a moderate seismic load of acceleration 0.25 g with very dense and soft rock soil properties (S_c). This study will involve the design of a structure according to two different seismic resisting systems. The 1st model is the special moment resisting frame system (SMRF) and the 2nd one is the shear wall system model (SWS). Then these models will undergo three wind speeds of exposure D, starting with low level of 45 mph, then with moderate level of 85 mph, finalizing with a high level of 125 mph. The results showed that wind load may dominate the structural design especially in shear wall system where high wind have more effect than seismic. On the other hand, the seismic effect in special moment resisting frame exceeds the high wind level effect and exceeds the seismic effect in shear wall system [6].

3 Methodology

In this paper, a 40-story reinforced concrete high-rise building will be constructed against a moderate to high seismic zone of acceleration 0.25 g. this building will resist the seismic load in three different resisting systems which are: the shear wall system (SWS), the special moment resisting frame system(SMRF), and the dual system (DS). After designing this building in these three systems, each of these systems undergoes three various wind speeds (low → 45 mph, moderate → 85 mph and high → 125 mph). This study continues the study present in the literature review of reference number 6.

This study has three essential objectives. It starts with the comparison of the seismic effect on the structure according to the three resisting systems. Then, comparing the effect of different wind speeds between the various resisting systems designed. Finally, the comparison between the wind effect and the seismic effect in each resisting system.

Figures 1, 2 and 3 show the plans of the two systems considered for the 40-story high-rise building. Table 1 shows the structural data for the two systems under study.

4 Results

A. *Total weight of the building*

The dual system is the heaviest of all having a total weight of 470236 kN with special moment resisting frame heavier than the shear wall system having weights of 461447 kN and 377573 kN respectively.

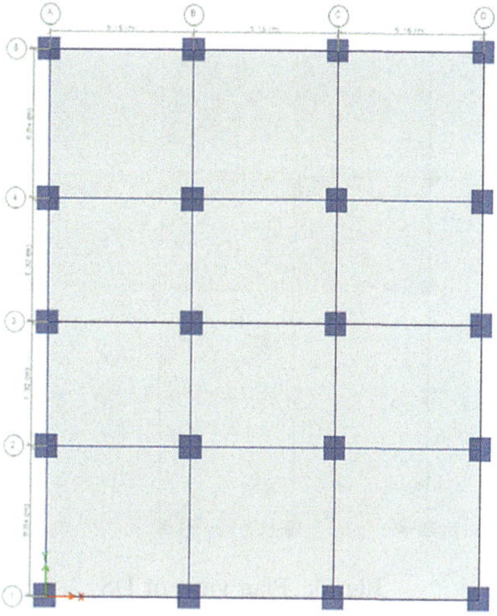

Fig. 1. Plan view of SMRF

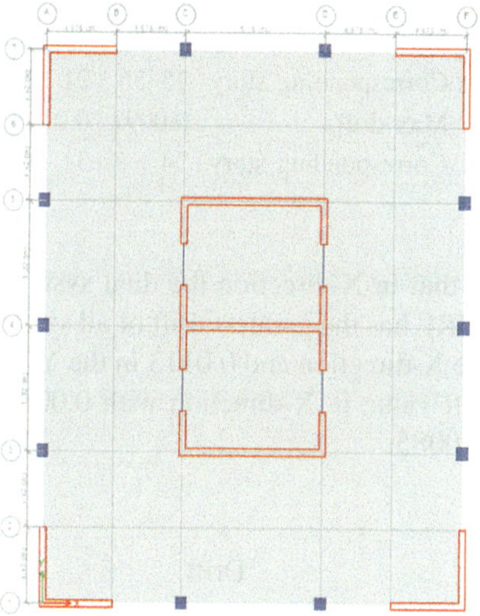

Fig. 2. Plan view of SWS

B. *Seismic Verification on Drift*

ETABS calculates the elastic story drift over the story height $\left(\frac{\Delta_s}{h}\right)$ so the inelastic story drift (Δ_m) must be calculated to compare it according to UBC 97 section 1630.9.2. he inelastic story drift must be less than or equal to 0.025 h when the period is less than 0.7 and must be less than or equal to 0.02 h when the period is greater than 0.7. The results of the max story drift for all the systems are under the limit which is 0.020 according to UBC 97 [7].

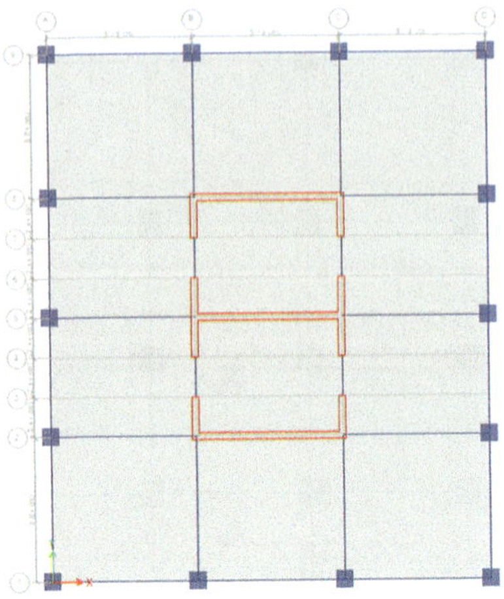

Fig. 3. Plan view of DS

Table 1. Inelastic drift

Direction	Max story drift	SWS	SMRF	DS
X	Max drift	0.0081	0.0130	0.0103
	Corresponding story	35–36	21	21
Y	Max drift	0.0093	0.0113	0.0085
	Corresponding story	24	31	21

It is noticed in Fig. 4 that in X-direction the dual system is in-between the SMRF and the SWS systems. SMRF has the greatest drift of all systems in both directions with drift equal to 0.0130 in the X-direction and 0.0113 in the Y-direction on the other hand, SWS has the smallest drift value in X-direction with 0.0083 and DS has the smallest drift in the Y-direction 0.0085.

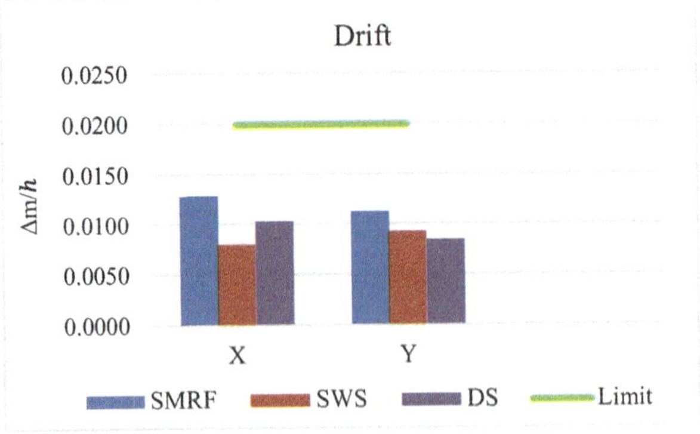

Fig. 4. Inelastic drift for the different systems

C. *Displacement*

Story displacement is the absolute value of displacement of the story under action of the lateral forces. There is a specific limit, according to UBC 97, for the displacement of a structure under the effect of wind. This limit is equal to $\frac{H}{500} = \frac{121{,}000\,\mathrm{mm}}{500} = 242$ mm (Figs. 5 and 6).

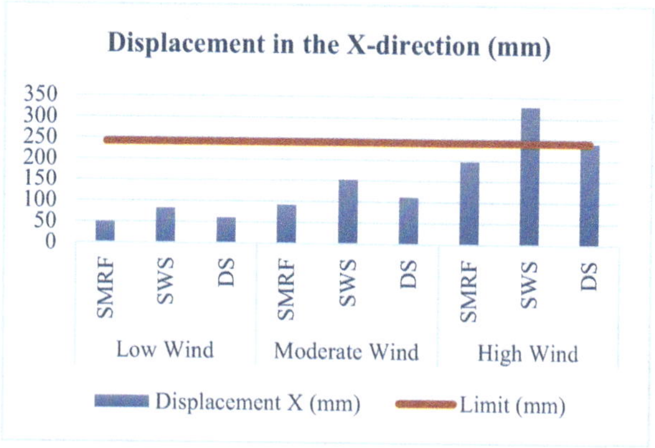

Fig. 5. Displacement in x-direction

This graph shows that for wind load cases of low and moderate, the displacement in the X-direction in all the resisting systems is under the limit (242 mm), having SWS the greatest of all followed by the DS and finally the SMRF. on the other side, for high wind SWS exceeds the limit by 72 mm. However, the SMRF and the DS are still under the limit even due to high wind speed, with the above graph displays the displacement of the 40-story building in the y-direction under the effect of low, moderate and high wind speeds for the resisting systems SWS, SMRF and DS. The displacement of the systems SMRF and DS are under the limit for all wind cases. However, the SWS is under the limit for low and moderate wind but exceeds the limit by 17 mm due to the effect of high wind level (Fig. 7).

D. *Base shear*

This graph and table above show that equality of the base shear between the systems in every wind speed, and the alteration in the base shear due to seismic between the systems. This is due to the way of calculation of each of the loads, where the formula of the base shear due to seismic includes variables that vary between the systems (the zone factor, the weight of the structure, the structural system coefficient and the near-by factor). On the other hand, the base shear formula due to wind load has no variables because of it is studied for the same building having same height, length and wind force. The base shear for low wind is 5,219 kN in the X-direction and 4,433 kN in the Y-direction, for moderate wind 9,653 kN and 8,198 kN respectively and for the high wind 20,876 kN and 17,730 kN respectively for all resisting systems.

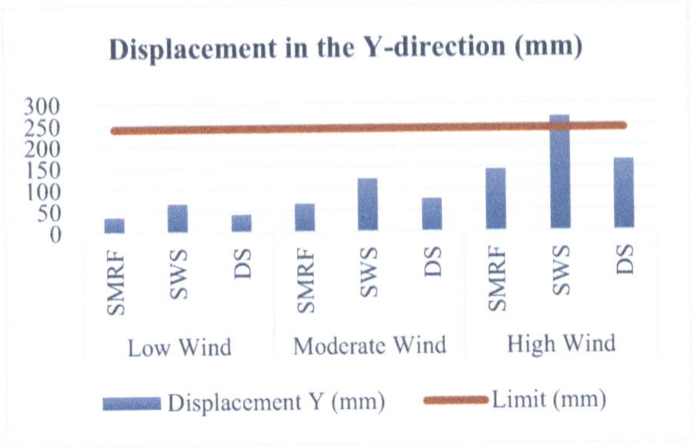

Fig. 6. Displacement in y-direction

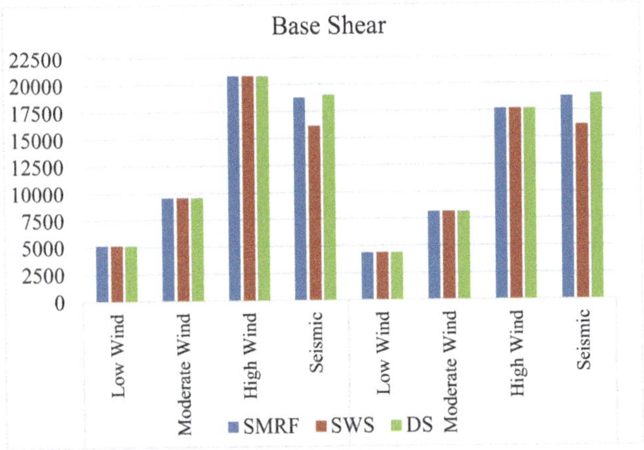

Fig. 7. Base shear

On the other hand, DS has a slightly larger base shear, due to seismic effect in both directions 19,118 kN, than the SMRF system with a base shear due to seismic of 18,855 kN. But the SWS has a base shear value of 16,219 kN which is the smallest value.

Moreover, the base shear due to low and moderate wind in the X-direction (5,219 kN and 9,653 kN) is less than that due to seismic (SWS: 16219 kN, SMRF: 18,855 kN and DS: 19,118 kN), which is by its turn less than the base shear due to high wind (20,875 kN). However, in the Y-direction the base shear caused by the seismic of SMRF and DS (18,855 kN and 19,118 kN respectively) has greater effect than all cases of wind even for the base shear due to high wind (17730 kN), but the base shear of seismic in SWS (16,219 kN) has greater effect than low and moderate wind yet smaller effect compared to the base shear of high wind (17,730 kN) (Fig. 8).

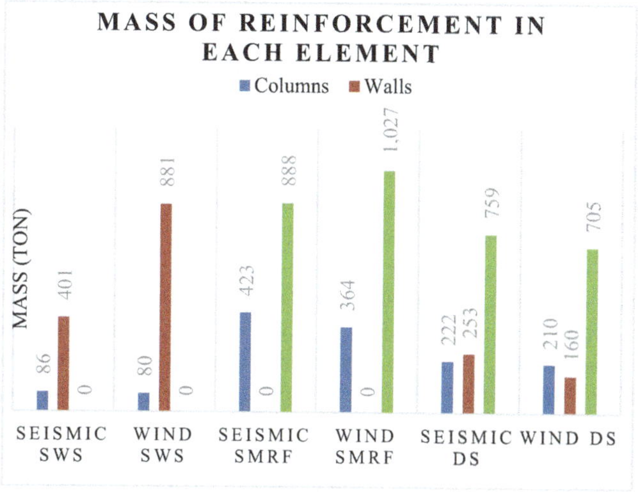

Fig. 8. Mass of reinforcement in each element (ton)

As a summary, by ordering the effect of base shear due to seismic and wind effect on the different resisting systems in an ascending order, the results are shown below:

- In the X-direction:

Low wind (all) < Moderate wind (all) < Seismic (SWS) < Seismic (SMRF) < Seismic (DS) < High wind (all).

- In the Y-direction:

Low wind (all) < Moderate wind (all) < Seismic (SWS) < High wind (all) < Seismic (SMRF) < Seismic (DS).

E. *Total Mass of Reinforcement*

This graph shows that:
For SWS, reinforcement of columns due to seismic load is slightly greater than that due to wind load of values 86 ton and 80 ton respectively. Alternatively, there is a huge difference in the wall sections were the reinforcement caused by wind load is more than double the reinforcement due to seismic load with the values of 881 ton and 401 ton respectively. Here, there is an essential information that needs to be highlighted. The first fourteen stories have the wind reinforcement dominating the design, which also includes an inadequate design due to shear and flexure upon exposure to high wind load for the first six stories. This is why we cannot consider the reinforcement mass value as an exact value. This indicates that shear wall system is not an adequate system for resisting high wind if it was designed for a seismic of 0.25 g. Also, it has the lowest mass which is not satisfactory for resisting wind load, in which the mass of the structure plays a big role in defending the building against wind load as mentioned in the literature.

For SMRF, reinforcement of columns due to seismic load surpasses that due to wind load having the following records 423 ton and 364 ton correspondingly. Hence, the value of reinforcement in beams increased to be 1,027 ton due to wind and it was

888 ton previously when the building was subjected to seismic load. Note that in this system there were no failure due to wind exposure neither for columns nor for the beams. It only requires greater area of steel reinforcement especially in the beams, where it needed to have more steel bars to defend the building against the extra load coming from the high wind load. In addition to that, most of the beams that needs to have more steel reinforcement areas are the exterior beams.

For DS, the reinforcement quantities due to seismic and due to wind loads are close to each other for the beams (759 ton and 705 ton respectively) and the columns (222 ton and 210 ton respectively), having the seismic reinforcement higher than the wind reinforcement. Conversely, for the shear walls located interiorly, the reinforcement of walls due to seismic noticeably exceeds that of wind load having the value of 253 ton and 160 ton correspondingly. Remark: For this system there is no failure for none of the structural elements. The dimensions of columns and beams have been enlarged to achieve the duality of the system as the UBC-97 code stated. This enlargement of the frames benefits the resistance of the structural against the wind load. The shear walls were placed interiorly, that's why it is not highly affected by the load coming from the wind.

Comparing the three systems, the beams reinforcement due to wind load in SMRF has the uppermost significance of reinforcement mass (1,027 ton). For walls, the mass of reinforcement for SWS due to wind load has the peak value (881 ton). Finally, the mass of seismic reinforcement in SMRF of columns has the maximum value of all systems' columns of value 423 ton (Fig. 9).

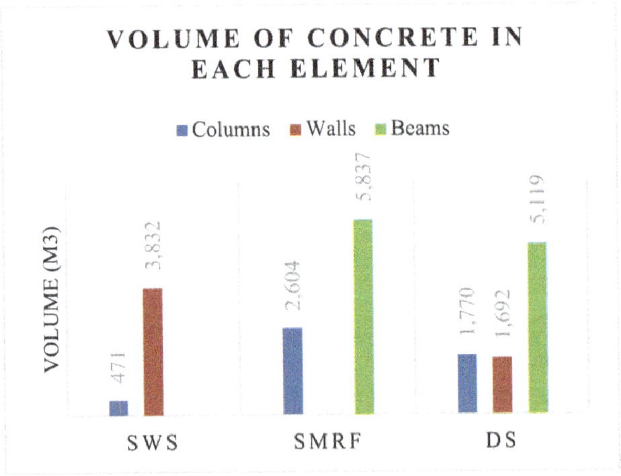

Fig. 9. Volume of concrete in each element (m^3)

F. *Total volume*

Volume of concrete in walls of SWS (3,832 m^3) is greater than that of DS (1,692 m^3) and is null in SMRF. SMRF has greater concrete volume than DS having the values of 5,837 m^3 and 5,119 m^3 correspondingly. Comparing the common element between the three systems, which is the columns, the following ascending order of the concrete volume is obtained, SWS with quantity of 471 m^3, DS having the value of 1,770 m^3 and finally the SMRF displaying a number of 2,604 m^3.

From these quantities, we can reach the total volume of concrete for each system. The following ascending order is resulted. SWS recording a total concrete volume of 4,303 m^3, then SMRF demonstrating a volume of 8,441 m^3 and the highest volume is documented by DS which is 8,582 m^3 (Fig. 10).

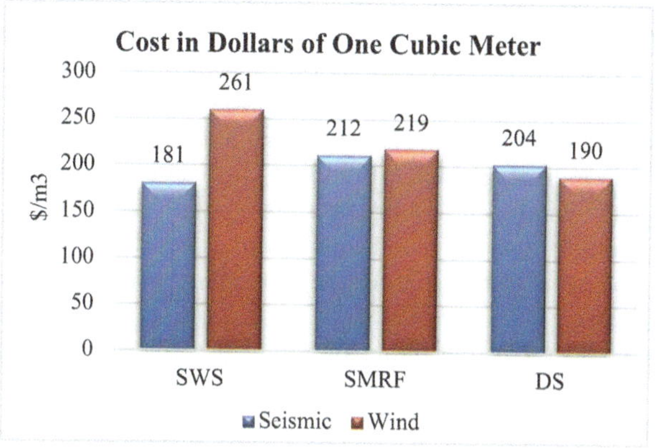

Fig. 10. Cost in one cubic meter ($/$m^3$)

G. *Cost*

For the SWS, the total cost of the building due to seismic load is equal to $780,792 which is cheaper than that due to wind which costs $1,122,148 with unsuitable dimensions for the wind load case.

For the SMRF, the total cost of the building is $1,787,859 when it is resisting the seismic load. This price is less than that of the building while resisting against wind load ($1,845,251).

For the DS, the total cost slightly differs between designing the building against the seismic load and the wind load. It costs $1,747,319 for designing against seismic load and $ 1,632,414 for designing against wind load.

SWS is the most economical system that can be used to resist a building against seismic load of charge 181 $/$m^3$ and the SMRF is the most expensive between the systems fighting against seismic load (212 $/$m^3$) having the DS as an in-between solution (204 $/$m^3$). Conversely, the most economical for a structure resisting wind load is the DS with a bill of 190 $/$m^3$ and the most expensive is the SWS which costs 261 $/$m^3$ there will be an extra cost when the suitable dimensions are used, while the SMRF is an in-between economical solution (219 $/$m^3$).

5 Conclusion

The vital specification for this study is that this high-rise building is a total regular building with no irregularities stated in UBC-97. The analysis above proved that high wind load of speed 125 mph can really dominate the design of seismic load of acceleration 0.25 g while low and moderate wind loads have no significant effect. Seismic and wind loads, each has an altered effect that varies with the category of the system used.

Beginning with the shear wall system, it is undoubtable that SWS is one of the preferable systems that can resist seismic load, and the analysis above proved these words are true. On the other side, regarding the failure of the walls when the building is subjected to high wind load, this indicates that high wind load affects the building more than seismic load does. This system is considered to be the worst system (between the three systems studied) to resist high wind load. This can be due to the lack of exterior barriers used against wind load and because this system has the lowest weight (mass) that plays a big role in resisting wind load (as mentioned in the literature [8]). This system inspires engineers to be use it for the economical price it covers when designed against seismic load.

The special moment resisting frame system is also a system that can resist seismic load. But regarding the results of the drift, the base shear, the story shear, the overturning moment and the reinforcement due to seismic load, SWS beats the SMRF. However, for high wind load, it is considered to be a vital system that resists this load. There was lack of failure in this system when it is subjected to high wind, but there was an increase in the reinforcement areas of beams due to wind effect implies that high wind load can govern seismic load also in this type of system. Most of the exterior frames have a higher reinforcement area when subjected to high wind load than to seismic load. This is because the wind load is exerted mainly on exterior frames and reduces as it goes to the interior (stated in the literature).

Finally, the dual system that combines both actions of the shear wall system and the special moment resisting frame system. This system is the most economical system to resist against wind load which encourage engineers to use it for high wind zones. It is also a system that surpasses SMRF to be used against seismic load and economically cheaper. This system has the frames which forms the exterior barriers against wind load and has the shear walls which can perfectly fight against seismic load.

References

1. Chopra, A.: Dynamics of structures: theory and applications to earthquake engineering
2. Heiza, K.M., Tayel, M.A.: Comparative study of the effects of wind, vol. 3 (2012)
3. Shashidhar, K.: Comparison of influence of wind and (2015)
4. Thilakarathna, S.N., Anwar, N., Norachan, P., Naja, F.A.: The effect of wind loads on the seismic performance of tall buildings. Athens J. Technol. Eng. **5**, September 2018
5. Aly, A.M., Abburu, S.: On the design of high-rise buildings for multihazard: fundamental differences between wind and earthquake demand. Shock Vibr. **2015**, 22 (2015)

6. Diab, A., Al-Rawi, Y.: Effect of Different Wind Speeds on a Seismically Designed High-Rise Building According to Different Resisting Systems, Lebanon (2019)
7. UBC (Uniform Building Code) (1997)
8. Ashok, A.: Difference between seismic and wind loads (2011)

Localized Effects Due to Aerodynamic Modifications in Buildings: A State-of-Art Review

Omkar Powar(✉) and Arul Jayachandran

Department of Civil Engineering, Indian Institute of Technology Madras, Chennai, India
omkarpowar.7@gmail.com, aruls@iitm.ac.in

Abstract. Assuring least spatial correlations is one of the basic principles for avoiding detrimental wind effects on high-rise structures. This is usually ensured in buildings by adopting aerodynamic modifications, which avoids peaking of the wind effects over different locations at the same instant of time, thereby reducing the wind load on the structure. There are many studies performed to quantify this aspect, but in most of these investigations, overall effects (like along-wind and across-wind overturning moment, power spectral densities and trajectories of various wind force coefficients) are focused, whereas the localized effects (like peak surface pressures) are generally not dealt in detail. In today's urban landscapes where the use of glass in building envelopes is very common, understanding the behavior of localized peak pressure distribution over building surface is of prime importance. Detailed quantification of these peak pressures is a pivotal issue for safe and economical designing of the glass envelopes. This paper provides a holistic review of existing literature based on both (a) wind tunnel tests and (b) computational fluid dynamic (CFD) analysis in the domain of aerodynamic modifications of high-rise buildings. The present paper highlights the typical limitation observed in most of these investigations wherein one member from each geometric modification family (like chamfering, recession or corner roundness) is compared with a member from other. Such investigation, although provide some qualitative information, are insufficient to address the practical issue of optimization faced by façade designer. The present paper establishes the need for benchmark studies, which encompasses all practically possible types of aerodynamic modifications that can be adopted in tall buildings. The paper also comprehensively presents a quantitative assessment of localized effects of wind on structural façades.

Keywords: Aerodynamic modifications · Computational fluid dynamics · Glass envelopes · Peak pressure · Wind tunnel test

1 Introduction

The primary function of a building is to protect the occupants from the harsh exterior climate and provide a safe and habitable interior environment. Today's high-rise building consists of two primary component system (a) structural load carrying system

H. Rodrigues et al. (Eds.): GeoMEast 2019, SUCI, pp. 98–127, 2020.
https://doi.org/10.1007/978-3-030-34216-6_8

(frames, bundled tubes, and so on) and (b) building envelopes which acts as a skin and physically covers the structural skeleton, these two come together to achieve the primary goal of the building structure stated above. There has been a continual evolution in the material used for the construction of these envelopes based on the tremendous technological advancements in the material research. Today most of the high-rise buildings use glass panels in their exterior façades, and the traditional brick masonry type of building envelopes are being replaced. As the material used has changed the critical forces governing the safe and economical designing of the building envelope system has also changed. In designing glazed, though the wind loads are dominant, overall wind-induced loads are not a governing factor. Here the localized distribution of pressure, location, and magnitude of the peak pressures are of great concern. The intensities and the nature of these pressures over façades have significant temporal and spatial variations. An estimate of the pressure to which the façade will be subjected can be done by three ways: (a) using the national code of practices, (b) performing wind tunnel tests and (c) performing numerical simulations using computational fluid dynamics (CFD). The first one being the easiest is mostly adopted by the design engineers. However, there are large numbers of approximations incorporated in this way of estimation. Also, most of the design standards provide the pressure coefficients only for specific basic plan shapes of a building and that too only for a selected set of building orientations. For any deviation from it, these design standards specify the use of specialist literature or go for the estimation by using wind tunnel testing. Today all the buildings will have some or other sort of aerodynamic modification from an aesthetic or structural point of view. In either case, the pressure distribution over these buildings will be different from that over the basic plan shapes of a building, which are generally provided in the design standards. This presents a challenge for ensuring safe, economical, and efficient designing of the glazed façades for such buildings. The present paper aims to encapsulate work done by various researchers and identify the research gap that still exists in this specific area of wind evaluation of buildings with aerodynamic modifications. This paper also provides basic guidelines for further research studies aspiring to contribute in the direction of the safe and sustainable designing the glazed façades for the buildings with aerodynamic modifications.

2 Wind Effects over Building

In general, all civil engineering structures behave as the bluff bodies immersed in the flow of air surrounding it. When a fluid (air) moves over a bluff body, it causes the flow separation and induces the so-called aerodynamic forces on the body. These aerodynamic forces can be categorized into two main types as (a) global wind effects and (b) localized wind effects depending upon its significance. Global wind effects are the effects, which are required to be considered when we have to design the structure as a whole. This category includes aerodynamic forces in along-wind direction (drag force), across-wind direction (lift force), base bending moment in both along-wind and across-wind directions; vortex induced vibration in the structure and the vortex shedding frequencies. These global phenomena play a crucial role in selecting the type of structural load resisting system, deciding the stiffness and the mass, which will be adequate to reduce the wind-induced responses. Similarly, localized wind effects are the effects which need to be considered when we have to design the individual elements like the glass panel in the curtain wall or the secondary structural systems like aluminum mullions and transom. This category of effects primarily revolves around the evaluation of various aspects of surface pressure distribution like the magnitude of peak pressure, correlations that exist in these peak pressures, locations and the width of the zones which theses peak surface pressures occupy.

Both types of wind-induced loads discussed above can be controlled passively by an appropriate choice of building shape and aerodynamic modifications. As described by Amin and Ahuja (2010), Mooneghi and Kargarmoakhar (2016) aerodynamic modifications can be classified into two main categories: (a) Macro aerodynamic modifications, which have considerable effects on the structural and architectural concept in design. (b) Micro aerodynamic modifications, which have negligible effects on the structural and architectural concept in design. Various types of commonly adopted aerodynamic modifications are categorized as shown in Figs. 1 and 2.

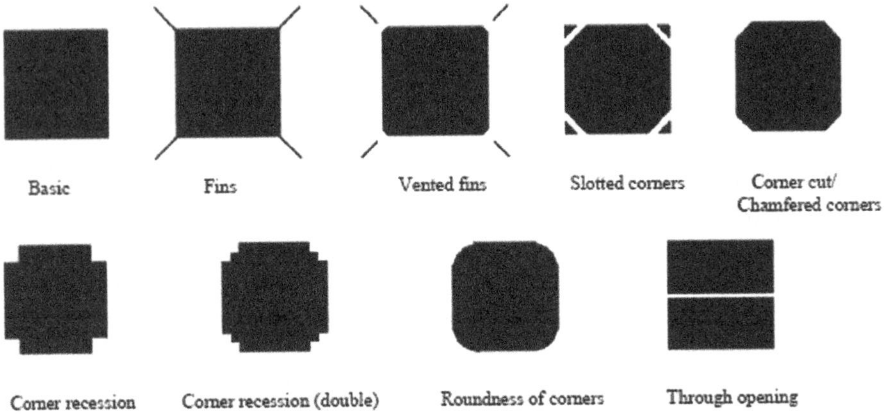

Fig. 1. Micro aerodynamic modifications (Source: Amin and Ahuja (2010))

3 Detailed Literature Review

3.1 Flow Behavior Around Buildings with Basic Plan Shapes

Circle, square, and rectangle belong to the highly used basic plan shapes of the building. For an understanding of the flow behavior and the factors governing the nature of the forces on these basic sections, there exists a large amount of literature out of which some are presented in this section. As there exists large portion of literature available on circular bodies that have already been well established and also reviewed by researchers like by Basu (1985) and Demartino and Ricciardelli (2017), in this paper we are reviewing only square and rectangular sections and their modified counterparts.

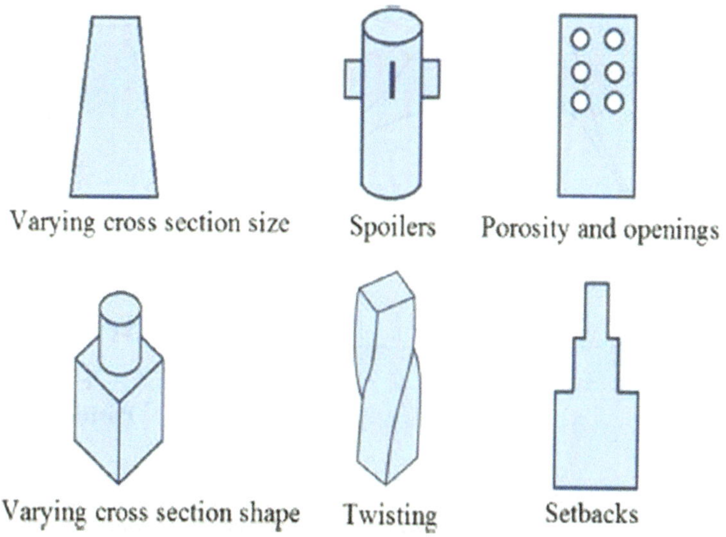

Fig. 2. Macro aerodynamic modifications (Source: Mooneghi and Kargarmoakhar (2016))

3.1.1 Effect of Turbulence and Reynolds Number

Vickery (1966) was amongst very early researchers to investigate the behavior of flow over the square cross sections in turbulent flow conditions. His primary focus was on assessment of the fluctuating lift force component and to understand the associated vortex shedding phenomenon. He also studied variation in the base pressure and pressure difference on the sidewall under the influence of turbulence, as shown in Fig. 3(a). He concluded that stream turbulence helps to increase the base pressure (reduction in suction) in the wake region and also reduces the fluctuating lift component by significant magnitude as high as 50%. Limitation of this study was that the experiments were performed only on the square section, and there was a need to perform similar experiments for rectangular sections with various side-ratios.

Hunt (1971) tried to elucidate the effect of shear and turbulence in the incoming flow over a building and concluded the effects as (a) wake is brought closer to the structure and (b) thickening of the free shear layer formed at the leading edges. His visualization studies have been beneficial in understanding the formation and interpretation of the surface streak lines over the building. He also explained the formation of the vortex at the front face of the building due to pilling up of layers in the incoming flow. This vortex sometimes called as the standing wave causes the down-flush at the front face and also a reason for the downward direction of the swirling flow on either side of the building.

Lee (1975) also performed an experimental investigation to understand the effect of turbulence on the drag and vortex formation phenomenon. For this purpose, he measured the surface pressure distribution on the prism with different level of turbulence intensity and concluded that along with the reduction in suction on the rear face the mean pressure distribution on the side face is also affected. In a uniform flow with turbulence intensity above 6.5%, a complete pressure recovery can be observed on the

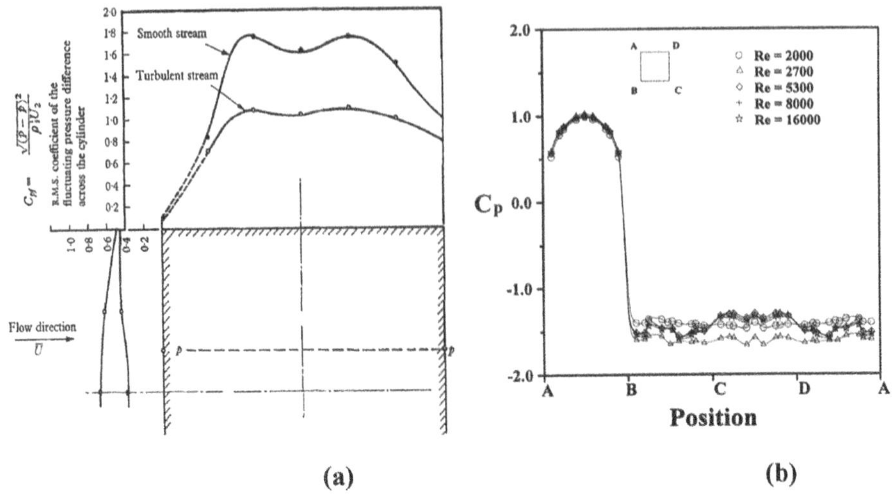

Fig. 3. Effect of (a) Turbulence and (b) Reynolds Number on the pressure distribution (Source: (a) Vickery (1966); (b) Chen and Liu (1999))

side faces of the prism which he linked with the effect of the thickening and inward inclination of the center line of the free shear layer originating at the leading edge. This thickening and inward inclination of free shear layer causes the intermittent reattachment of flow on either of the side of the prism and pushes formation of vortices on the downside in the wake ultimately resulting into increased base pressure (reduced suction) in the wake and hence reduced drag on the prism. Limitation of this study can be pointed out that though the pressure distribution was measured and studied, the focus was on the understanding of the global phenomenon.

Huot et al. (1986) extended the study similar to Lee (1975) and confirmed that the mean pressure distribution in the zone upstream of the separation of flow is not affected by the turbulence from incoming flow whereas the pressure distribution on the downstream of separation point is strongly influenced by the characteristics of free shear layers. They confirmed that base pressure is related to the curvature of the free shear layer, stronger the curvature higher is the suction in the wake. Effect of initial turbulence is to enhance the entrainment of air from the recirculation zone, increasing the outward curvature of free shear bringing the vortices close to the rear of section and hence increasing the suction in the wake. However, the addition of turbulence further helps the shear layer to reattach on the sidewall of the section and formation of two new separation shear layers originating at the trailing edge. These separation layers are comparatively weak and hence increasing base pressure and reducing the drag on the section.

Nakamura and Ohya (1984) studied the effect of turbulence over the rectangular section and concluded that for small turbulence the rate of growth of free shear layer is increased due to increased mixing whereas for more considerable turbulence the span correlation is reduced which weakens the formation of vortices in the wake region.

Kareem and Cermak (1984) and Kareem (1990) performed a thorough investigation to access the effect of turbulence in the incident flow over the fluctuating pressure of a square cylinder and rectangular cylinder (B/A = 0.67 and 1.5) respectively. Two sets of approach flows (resembling open and urban terrain) were simulated in the wind tunnel for testing. They were able to pinpoint some critical conclusions as; (a) Turbulence in the flow encourages early reattachment of flow and corresponding pressure recovery over the side face of the section hence reducing the lift coefficient. (b) Study of pressure fluctuations showed that comparatively higher correlations exist in chord-wise (along with the flow) direction for the tap locations which are far from the top and bottom portion of the cylinder. Similarly, a good correlation in pressure fluctuations can be seen in the separation bubble region along the span-wise (along with the height) direction. Increase in turbulence weakens both chord-wise and span-wise correlations in the fluctuating pressure over the side face. (c) Nonhomogeneous nature of the fluctuating pressure hints that not only separation distance but also the relative location of it is of prime importance in the understanding of flow characteristics over the building. They emphasized that the estimation of point-to-point pressure (both mean and fluctuating component) are essential from the perspective of designing individual elements like cladding, its attachments, and architectural add-ons.

Vortex shedding frequencies of the rectangular section with side ratio (B/A = 1 to 4) over a vast range of Reynolds Number (Re) ($70 < Re < 2 \times 10^4$) were evaluated by Okajima (1982). He initially stated that "In the case of a sharp-edged body, like a rectangular cylinder, separation is fixed at the leading edge and the aerodynamic characteristics are said to be relatively insensitive to Reynolds number." However, later he also mentions that for flow with extremely low Reynolds number, the flow may immediately reattach to the side face and then finally separates at the trailing edge.

Chen and Liu (1999) performed experiments on square cylinder for a range of Reynolds Number (Re = 2000 to 21000) to understand its effect on vortex-shedding frequencies and surface pressures coefficients. They found that pressure distribution on the windward face is almost insensitive to the Reynolds Number (Re) whereas there is a slight increase in negative mean pressure on the side and leeward face as shown in Fig. 3(b). For lower Reynolds numbers (such as Re. = 2000 to 3300), the maximum Strouhal number occurs at a relatively higher angle of 17°, whereas when the Reynolds number ($Re \geq 5300$) the Strouhal number shows a similar trend of rapid rise in Strouhal number (hence minimum drag coefficient) but at a lower angle of around 13°. They also confirmed the claim made by other researchers that the phenomenon of the rise in Strouhal number is due to reattachment of the separated shear layer which takes place due to strong pressure recovery on side face with increase in the angle of incidence.

Norberg (1993) and Steggel (1998) performed experimental investigation and also compiled the work done by most researchers like Okajima (1982), Igarishi (1985) along with the numerical studies of Davis and Moore (1982), Franke et al. (1990), Okajima (1990) and Ohya et al. (1992). Though there existed considerable scatter in the result of various researchers, Steggel (1998) comprehended the Reynolds number effect over Strouhal number as (a) for square section (B/A = 1.0) the Strouhal number could be considered as almost continuous and constant for higher Re values (Re > 1000). (b) Whereas in the case of sections with (B/A = 3.0) Strouhal number can be considered as continuous and constant above Re > 5000. He also concluded that for flow below the critical Reynolds number flow separates at the leading edge but again come back and gets attached to the side face of rectangle weakening the vortices formed in the wake and causing a discontinuity in the Strouhal number. For square section, the after-body length is too short for reattachment whereas for rectangular sections with (B/A > 4.0) the after-body length available guarantees that the flow will reattach on the side face and thus making both these cases independent of the Reynolds number. All the above summary given by Steggel (1998) comes with the caution that most of these studies to evaluate the Reynolds number effect have the free stream turbulence level as low as 0.5% turbulence intensity and hence should be adopted carefully in practical application.

3.1.2 Effect of Side Ratio

Bearman and Trueman (1972) experimentally investigated over the rectangular cylinder of side ratio (B/A) varying between 0.2 to 1.2 to evaluate its effect over the base pressure and corresponding drag coefficient. Prior to Bearman and Trueman (1972), it was believed that all bluff bodies are generally subjected to a uniform drag coefficient of approximately 2.0. This postulation was based on the data available till then, which primarily comprised experimental results on either flat plate or long cylinder with a square cross-section. Hence the experimental results of Bearman and Trueman (1972) came as a surprise which said that for a rectangular section with side ratio (B/A) of 0.62 the maximum drag coefficient of 2.94 could be observed. Further examination showed that there exists same and almost uniform pressure distribution on the rear face for sections with size ratio (B/A = of 0.2 and 1.0) whereas a high suction was observed around the center of the rear face for the section with a side ratio of 0.62, as shown in Fig. 4(a). He justified this increase suction as the effect of the interaction of two vortices formed at the two side faces of a rectangle, resulting in enhanced negative base pressure in the wake region. In this study though the surface pressure distribution was studied, the focus was still to understand the global phenomenon of drag and vortex formation in the wake.

Akins and Cermak (1976) performed a systematic experimental investigation to evaluate the pressure distribution over the isolated building under a set of realistic boundary layer flow conditions. Akins showed that the side ratio (B/A) of section adopted is primary factor governing the mean pressure distribution and other factors like aspect ratio (H/B) or the approach flow conditions can condense to obtain one set of mean pressure coefficients for the given orientation of the section. He also confirmed that the pressure fluctuations acting over the side and rear face of the building are primarily governed by the flow characteristic around the building rather than approach flow conditions. In general, it can be said that there exists a very weak correlation between the pressure fluctuations on all sides of building and velocity fluctuations in approach. He was also able to identify the location of reattachment of flow, which occurs at the point associated with the local maximum RMS pressure on the side face of the section.

Laneville and Yong (1983) performed an experimental study with flow visualization technique to obtain insights to the physical reasoning behind the increase in drag at the particular side ratio of (B/A = 0.62) as demonstrated by Bearman and Trueman

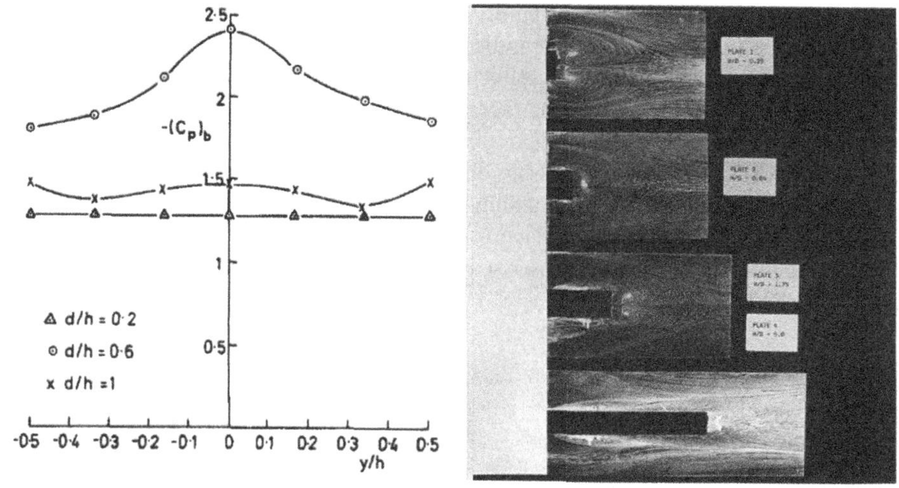

(a) (b)

Fig. 4. Effect of side ratio over (a) Base pressure and (b) Formation of vortices (Source: (a) Bearman and Trueman (1972); (b) Laneville and Yong (1983))

(1972). As shown in Fig. 4(b), he was able to visualize the formation of two separate vortices for all the side ratio (B/A = 0.25 to 5.0) from either side of the rectangle. For small side-ratio (B/A < 0.5) the after-body length is too small to interact with the flow separated at the leading edge. When the side ratio increases the thickness of free shear at the leading edge also increases, and the separation bubble grows, which influences the vortices to form close to the rear face of the rectangle. Above phenomenon leads to the increase in suction in the base pressure in the wake region, ultimately resulting in the higher drag coefficient. With further increase in side-ratio (B/A > 0.6 to 0.7) the separation bubble is considerably away from the trailing edge reducing its potential to influence the formation of vortices in the wake region. Thus, the vortices start to move downward away from the section reducing drag again.

Igarashi (1985) executed an experimental investigation (both wind tunnel and smoke flow visualization) to ascertain the effect of the side ratio (B/A) from 0.1 to 4.0 over surface pressure distribution (mean and fluctuating) on all the faces, as shown in Fig. 5. He summarized his observations as; (a) For side ratio 0.5 the negative mean pressure (suction) gradually increases from leading edge to trailing edge whereas the negative mean pressure on rear face is almost uniform and have a slightly higher value than side face. Fluctuating pressure at this side ratio is also found to be considerably large. (b) For side ratio of 1.0, the negative mean pressure on side face has a higher value than that on the rear face, and also there is a discontinuity in values near the trailing edges. Here though the fluctuating pressure values are lowered on rear face the fluctuating pressure values on side face remain comparatively high. (c) For further increase in side ratio up to the critical value of around 2.5 to 2.8, the negative mean pressure (suction) on the rear face is reduced. (d) For side ratio 2.0 fluctuating pressure for side face is comparatively large but for side ratio 2.5 it is brought down and remains almost same as that for the rear face indicating the reattachment on side face and weak vortex formation in the wake. (e) After reattachment for side ratio 3.0 and 4.0, the negative mean pressure on the rear face and near trailing edge on side face reduces gradually whereas the fluctuating pressure is increased in these regions. Thus, the critical ratio of 2.8 divides the pressure distribution into two distinct patterns. For side ratio smaller that critical value fluctuating pressure tends to rise with the increase in the value of negative mean pressure (suction) whereas when side ratio is higher than the critical value the fluctuating pressure rise in the region where there is a reduction in the value of negative mean pressure.

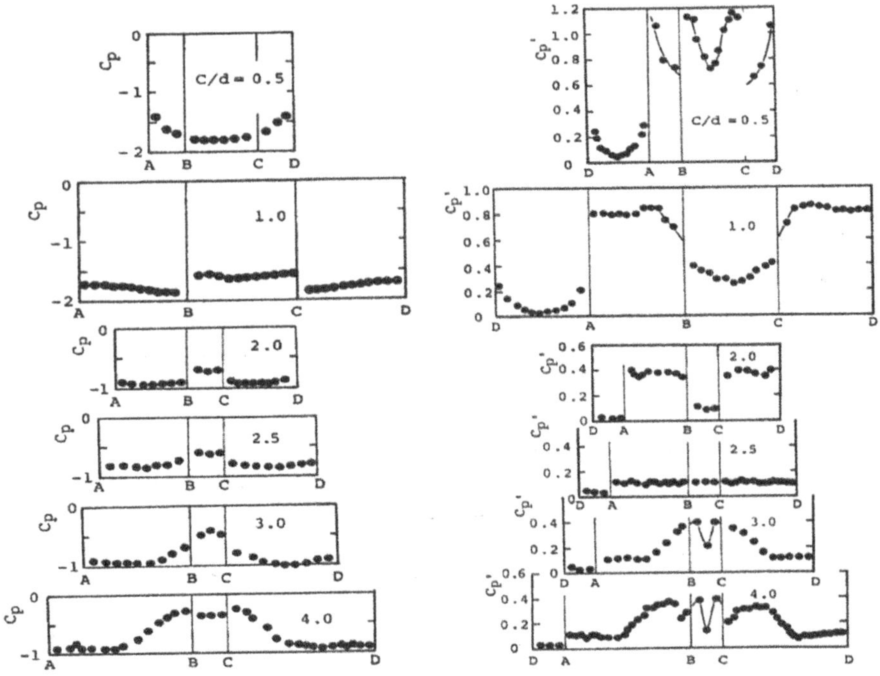

(a) (b)

Fig. 5. Effect of side ratio on (a) Mean pressure distribution and (b) Fluctuating pressure distribution (Source: Igarashi (1985))

Okajima (1990) performed a numerical investigation to gain insights to the flow characteristics of flow with high Re number over the rectangular section with the side ratio (B/A) varying from 0.6 to 8.0. He observed that there exist critical side ratios with a value of 2.8 and 6.0, which demarcated by an abrupt change in Strouhal number. Around the side ratio of 2.8 the flow almost changes from fully separated flow to alternately reattaching flow altering the value of Strouhal number whereas when the side ratio is as high as 6.0 the separation bubble is almost attached on the sides of section, and the vortex is shed at comparatively higher frequency again increasing the value of Strouhal number. He also mentioned that for the sections with higher side ratios, the drag coefficient is reduced gradually and a maximum of drag exists for the section with a side ratio of 0.6.

Norberg (1993) also extended the study to evaluate the effect of both side ratio (B/A = 1, 1.62, 2.5) and 3and inclination (0° to 90°) over the rectangular cylinder, key difference here was that he also captured the mean surface pressures along the centerline of these sections. Similar to Bearman and Trueman (1972) he found that the value of drag coefficient for the section with (B/A = 0.6) to be maximum as shown in Fig. 6(b), and the pressure distribution on the rear face becomes non- uniform with very high suction near the center of the rear face. Physical reasoning for this increase was obtained through flow visualization as an increase in the base cavity due to very powerful vortices shedding and strong curvature of the free shear layers for this side ratio.

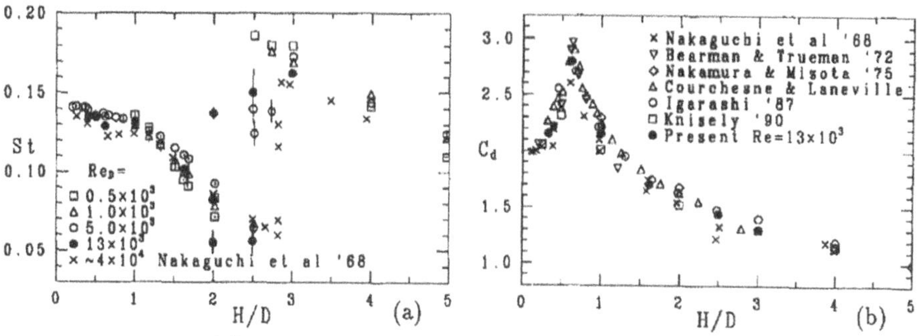

(a) (b)

Fig. 6. Effect of side ratio on (a) Strouhal Number and (b) Drag coefficient (Source: Norberg (1993))

Ohya (1994) studied detail and tried to fine-tune the effect of side ratios (0.4, 0.5 and 0.6) with specific focus over the base pressure. He observed that though most have researchers confirmed that the side ratio of 0.6 is the one with the highest base pressure and corresponding highest drag occur, the side ratio of 0.5 also has a distinct transition feature. It is the side ratio, which shows the two different base pressure values, high and low, both corresponding to two different sets of flows at an irregular interval of time gaps.

Bruno et al. (2010) performed a computation study over a rectangle with side ratio (B/A = 5.0) under high Re number and turbulent flow conditions. As shown in Fig. 7, they identified four crucial zones on the surface of the rectangular section, and their corresponding distinct characteristics based on the analysis of 2D instantaneous flow field, mean surface pressure, and fluctuating pressure. These zones are (a) vortex shedding zone, which starts from the leading edge where flow separation occurs and identifies by a plateau of mean suction and low fluctuating pressure value. (b) Vortexcoalescing zone, where the coalescence of the primary vortices happens and mean suction attains its highest value, and this zone extends up to the point where the vertical component of reverse flow induced by vortex exist. (c) Pressure recovery zone, where most of the mean suction present on the side face gets recovers, and it is also marked with the occurrence of maximum fluctuating pressure. (d) Mean reattachment flow zone where the actual reattachment of the separated flow happens and is characterized by a plateau of mean suction. More studies of this type performing detailed analysis, especially over localized effects and flow visualization for various side ratios are required for safe and economical designing of the glass façades in the future.

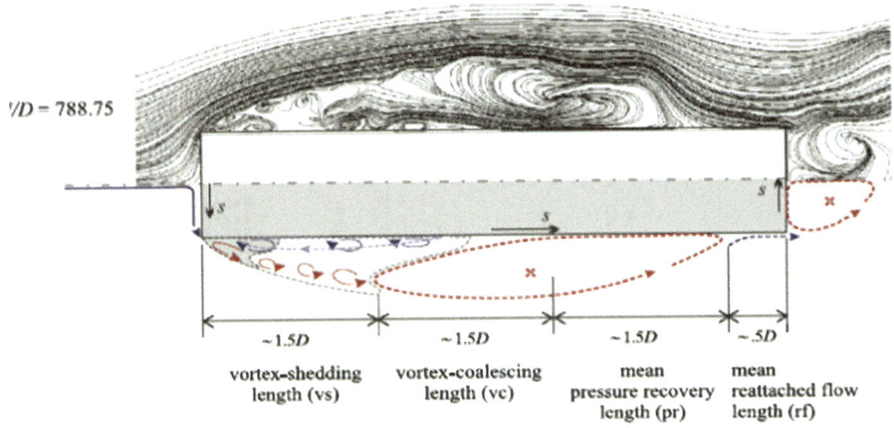

Fig. 7. Instantaneous flow field on one side and schematic representation of identified zones on another side of the rectangle (Source: Bruno et al. (2010))

3.1.3 Effect of Orientation

Robertson et al. (1978) investigated the square prism in uniform flow to get insights into the separation-reattachment phenomenon. Mean and fluctuating pressure values were measured for orientation (−50° to 90°) and the incoming flows with and without turbulence. The term "separation bubble" was defined as the tapered flow initiating at the leading corner up to the side wall reattachment of flow and the changes in its characteristics with inclination were studied. For flow conditions with low turbulence levels, the inclination greater than 14° would result in the flow reattachment whereas the reattachment was confirmed at only 9° inclination for the turbulent flows. The length of the separation zone and the magnitude of pressure variations in it were found to be inversely proportional on the inclination after 14° inclination and would become almost zero for 45° of inclination. Location of reattachment was identified in the zone between the peaks of mean and fluctuating (RMS) pressure measured on the side wall. Turbulence in the flow would encourage the reattachment phenomenon but also push its location near the rear edge.

Igarashi (1984) took the initiative to perform a detailed investigation on characteristics of flow around the square prism at an angle of attack 0° to 45° in the range of subcritical the Reynolds Number. Based on the type of the flow separation and reattachments pattern he identified four regimes flow according to angle of attack (α) as (a) Perfect Separation type (Symmetric flow type): $0° \leq \alpha \leq 5°$, (b) Perfect Separation type (Unsymmetrical flow type): $5° \leq \alpha \leq 13°$, (c) Reattachment flow type.: $14°–15° \leq \alpha \leq 35°$, (d) Wedge flow type: $35° \leq \alpha \leq 45°$. Through analysis of the effect of angle of attack (α) on vortex shedding frequency (f) concerning the above-mentioned flow types are discussed in detail. Igarashi (1984) also discusses the effect of angle of attack (α) on mean pressure and fluctuating pressure distribution around the prism with the help of Fig. 8. He states that with an increase in the angle of attack (α), the Stagnation point on the front face (DA) approaches the leading edge (A). Moreover, for (α) $\geq$ 15° shear layer separated from the leading-edge re-attaches on the lower face AB, X_R and X_S (maximum and minimum C_p value locations) are observed to shift towards the leading-edge A and also the values of maximum and minimum C_p are also increased with increase in (α). He also explains about the formation of separation bubble due to reattaching of the shear layer at an angle of attack (α) $\geq$ 15° and mentions that size (length and height) of bubble decrease with increase in (α). Details of the fluctuating pressure which are observed to be high in the separation bubble region are also discussed in detail.

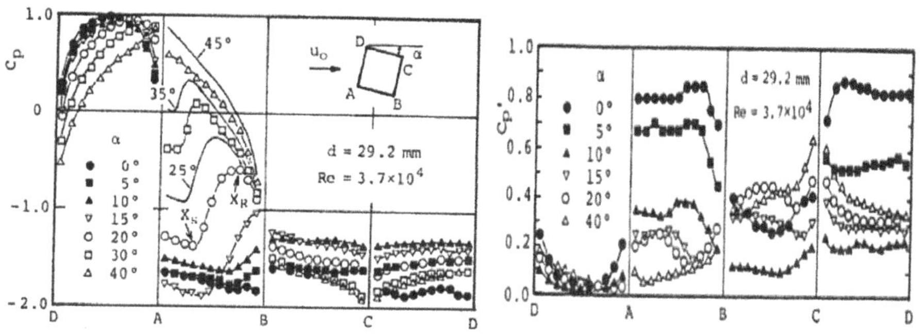

(a) (b)

Fig. 8. Effect of orientation on (a) Mean pressure distribution and (b) Fluctuating pressure distribution (Source: Igarashi (1984))

Knisely (1990) performed the experimental investigation to ascertain both the effect of side ratio (B/A 0.04 to 1.0) and inclination (0° to 90°) over the rectangular section. He compared his results with some of the prominent researchers at that time and summarized as there is a general tendency of a sudden increase in the value of Strouhal number at lower inclinations, i.e., around 12° to 15°, which also accompanies with a drop in drag and lift coefficient. He corroborated with the finding of Igarashi (1984) that this inclination of a sudden increase in Strouhal number can be considered as the effect of reattachment of the free shear layers over the side face of the section. Further, he also concluded that this inclination is dependent on the side ratio (B/A) and also for a given side ratio there exist an optimum inclination angle which is most efficient from load-bearing considerations of the structure. Even in this paper also a typical limitation was observed that the focus of the study was on the global aspects only and localized effects due to surface pressure distribution were not give importance.

Dutta et al. (2003) worked on similar lines to evaluate the effect of the inclination over the square section. He compared his work with the researchers like Norberg (1993) showed that there is a sharp increase in Strouhal numbers accompanied by a drop in drag coefficient for initial inclinations from 0° to 22.5°. To explain the physical reason for this they performed flow visualization sturdy and found the separation for 0° occur at the leading edge and as the inclination is increased the separation point moves downstream helping the pressure recovery and reattachment of flow to occur easily. This phenomenon results in lower drag and higher Strouhal numbers.

Lipecki et al. (2012) performed a experimental investigation over five rectangular cylinder of finite height to understand the effects of various factors such as side ratio (B/A = 2.0 and 4.0), orientation (0° to 90° at interval of 15°), size of the models adopted, and six wind profiles with varying turbulence levels. Wind tunnel results were presented in the form of spatial distribution (contour maps) of mean surface pressure and circumferential and vertical pressure distribution for various combinations of above-effecting factors. Some of the primary conclusions being; (a) the highest values of positive mean pressure occur in the middle of the windward wall at the height of about 80–90% of the model independently from the wind structure. (b) Contour lines connecting points of equal values of mean pressure coefficients for angles 15° to 75° are arranged in parallel to the vertical edges of the model and show a peculiar pattern of reduction in magnitude near the top due to 3D flow around the free-end. (c) Vertical distributions of mean wind pressure coefficient show a clear dependence on the mean wind speed profile. (d) Mean pressure distribution pattern remains almost similar for both the side ratios considered, but the magnitudes are affected, more studies are required for getting insights on this aspect.

3.2 Flow Behavior Around Buildings with Aerodynamic Modifications

Davenport (1971) was well ahead of his time and initiated the investigation of flow over the buildings with aerodynamic modifications. He performed the wind tunnel evaluation of six tall building models (square, rectangle, triangle, circle, recessed-square, recessed-triangle) and measured the mean and the fluctuating responses at the roof height in both along and across wind directions. He stated that significant effects were observed in the peak displacements at roof height which can be viewed from both sides as aerodynamic penalties associated with the use of the different shape of buildings or the way to improve the performance by adopting proper shape. From the Fig. 9, it can also be said adopting corner recession in square model results into reduction in peak displacement. However, this paper also suffers from the common drawback that only a single member from each modification family was selected for comparison.

Kareem (1983) elaborated the ways of mitigation of wind effects over a tall building in two ways (1) Passive Mitigation and (2) Active Mitigation. Though he primarily focused on the active mitigation method like the use of mass dampers to

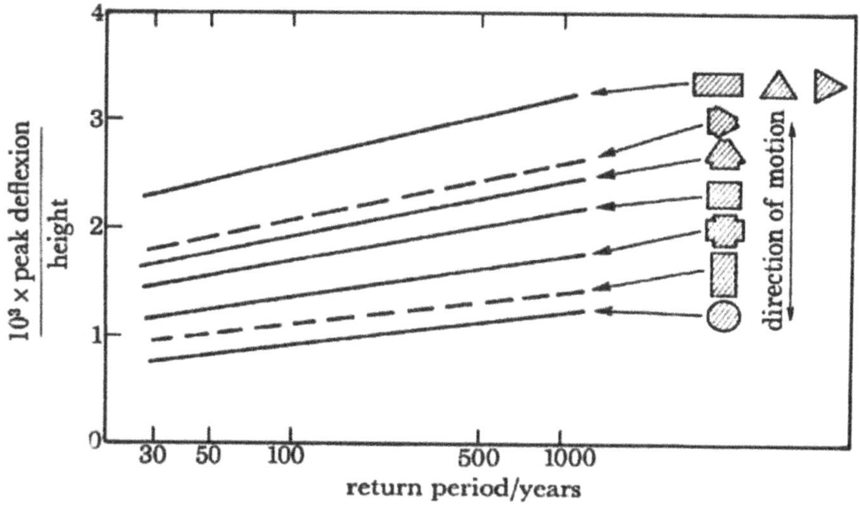

Fig. 9. Comparisons of peak deflection at roof level for various building configurations with and without modification (Source: Davenport (1971))

reduce the wind-induced motions in the building he also discussed various aerodynamic modifications, their effectiveness and the theoretical basis on which they work. Kareem (1983) stated that use of small vanes at the building corner provides the passage of air through them directly to the side face of building reduce suction under the separated shear layers which generally form at the leading edge of the building However, an additional drag will also be introduced by these vanes resulting into enhanced along wind response.

Detailed wind tunnel investigation was performed by Kwok and Bailey (1987) to evaluate the effect of (a) fins, (b) vented fins and (c) slotted corners over a the along and across wind responses over a square cylinder with an aspect ratio of (H/B = 9.0). The normalized mean and standard deviation of roof displacements were measured and concluded that there is an overall increase in along-wind response in both fins and vented fins than that of the square prism, and the reason for this was identified as an increase in projected area. Whereas on the other hand, air infiltrating through the slotted passage helped to alter the pressure distribution and as a result, slotted corners were found to be effective in reducing both the mean and standard deviation of the along and across wind response.

Kwok (1988), Kwok et al. (1988) further investigated effects of aerodynamic modification like (a) slotted corners, (b) horizontal slots (through opening slits) at the half and 2/3rd elevation level and (c) chamfered corners over a standard rectangular benchmark building (CAARC) model as shown in Fig. 10. Wind tunnel experiments were performed in two boundary layer profiles (open terrain and sub-urban terrain) to measure the mean and standard deviations of the along and across wind displacements at roof level of the model. They found that the mean and standard deviations of the responses were proportional to the reduced velocities. Chamfered corners were found

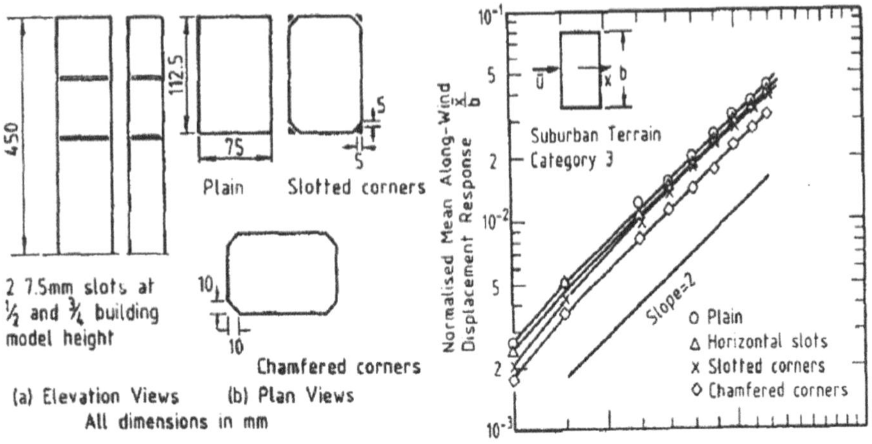

Fig. 10. Comparisons of along-wind peak deflection at roof level for CAARC building model with and without modification (Source: Kwok (1988))

very effective and reduced almost 40% of the responses as compared to the building without modifications.

Hayashida and Iwasa (1990) experimentally measured the aerodynamic forces and displacement responses at the roof level for the 600 m tall building. Total of eight rigid models (base section as circle, square and triangle with and without aerodynamic modifications) have been studied. Results showed that rounding of corners to be more effective than the corner recession to reduce the cross-wind displacement responses of building with square section. The comparison of all the models was also performed, but as shown in the figure and triangular base models were found to perform better than the square section models in general.

Dutton and Isyumov (1990) studied the effects of different vertical gap widths (in both along and across wind directions) over the crosswind response of a square cross-section tall building. They found these modifications to be effective in disrupting the formation of the vortex shedding in the wake and hence reducing vortex- shedding induced crosswind forces and displacement responses of the tall buildings. Effectiveness varies with the proportion of gap width. For gap width in the range of about 4% to 5% a substantial reduction is observed whereas there is again some rise in the response in the range of 10% to 15% the physical reason to which sated as unknown. Such studies on the effect of the proportioning have shown that sometimes there may be adverse effects of aerodynamic modifications with specific proportions, and this aspect needs more attention from researchers.

Okamoto and Uemura (1991) studied the effect of rounding off all the corners of the cube subjected to the boundary layer flow through flow visualization and wind tunnel testing. He observed the formation of two types of vortices which are generally seen around the flow in circular and rectangular cylinder namely; (a) horseshoe vortices forming between the front face of cylinder and ground which further gets down washed along the side-wall of section and (b) Arch vortices which forms behind the cylinder.

They found that the drag coefficient reduces with increase in rounding of corners, and surface pressure distribution approaches that of the circular cylinder.

Jamieson et al. (1992) studied the effect of chamfered, rounded and recessed corner (approximately 10% and 30% of smaller side) on the pressure distribution over the CAARC standard tall building model. They observed that peak positive Cp for each shape was about +2.0, whereas the variation in the peak negative was significant for different modifications and proportions. For the base model without modification peak suction (Cp = −4.1) occurred close to trailing edges on the shorter side for the orientation perpendicular to the face. Also, broad bands of suction (Cp = −3.5) occur on the same faces about 1/3 distance back from the leading edges. In case of recessed corners the peak negative Cp occurred to be at a similar location near trailing edge but at the orientation of around 31° and the magnitude was slightly reduced. Effect of chamfering was to reduce the peak suctions on the trailing edge on the side face, and this modification resulted in weakest peak suctions. Rounded corners were observed to produce strong peak suctions near the center of the windward curves and had the adverse effect as the suction were higher than that of the base model without modification. The primary limitation of this study is that measurement of the pressure distribution was only performed at a single elevation level (2/3rd height of the building), and the variations along the height were not evaluated.

Miyashita et al. (1993) performed a detailed wind tunnel evaluation is performed for six square building models having an aspect ratio (H/B = 6.0) with the aerodynamic modification of chamfering (10%), single recession (10%) and opening or gap at different elevation levels. Along and across wind forces along with the overturning moments at the base of all models were measured for the inclinations varying from 0° to 45° at an interval of 5°. It was observed that for all the types of modifications considered the for across wind force fluctuations were reduced than that of a base square model for the orientation of 0° and the model with an opening in both X and Y directions together was most effective. Comparatively, a weak correlation exists for the wind forces in X and Y direction for the low angle of inclinations whereas good correlation was observed for the higher inclinations from (20° to 45°) in the region of vortex shedding frequency.

Lou et al. (1994) performed both quantitative (wind tunnel) and qualitative (water tunnel flow visualization) to analyze the effect of after body shape of the section with and without inclination. He studied four cross sections with the same upstream face, whereas the side face and the rear face of the base square model were modified to obtain two trapezium sections and one triangular section. He pointed out that the primary effect of the variation of after body shape occurs on the proximity of the free shear layers among themselves and with their side faces, which ultimately governs the flow phenomenon around the section. This proximity not only influences the mean and fluctuating pressure on the side face of sections it also influences the inward or outward inclination of the free shear layers, which in turn governs the vortex formation length and reattachment phenomenon.

Investigation performed by Surry and Djakovich (1995) concentrated on the effect of geometry and the turbulence over the pressure fluctuations over the side face of the building. He has performed detailed wind tunnel and flow visualization studies over four building configurations (Square, two diamond shape models and square model with the single recession) and presented results in form contours of maximum, mean and minimum pressure coefficients (Cp) over the building for three terrain categories. Their results can be summarized as (a) Peak positive and positive mean Cp values exist near the stagnation point on the windward side and have the general tendency to move to higher elevation along with an increase in the terrain roughness. (b) For square model Peak negative Cp exist along the leading edge and mostly in the lower 2/3rd region in contrast to the mean negative Cp which exist along the leading edge but typically in the top region. (c) Comparison of the Peak negative Cp for different configuration did not show any unusual variation in magnitude, but the location and the critical angle at which the peaks occur are definitely affected. Peak negative Cp for models with diamond cross-section lies in the acute corner regions near the base whereas for the section with single recession peak negative Cp exist along the vertical centerline of the side face of the building. He made a very bold and farfetched conclusive statement that "details of corner geometry do not significantly affect the general magnitudes or patterns of peak suctions, but rather the wind angle at which they occur." However, further, he compensated by stating that the peak suction phenomenon and its affecting factors are not yet fully established with firm cause-effects relationship and need to be studied further.

Kawai (1998) investigated instabilities such as vortex induced excitation and galloping oscillation of square and tall rectangular building. He performed aero-elastic wind tunnel testing over models with modifications like single recession, chamfering and corner rounding with the proportions varying from 5% to 25% of the width of the section. They found that the small proportion (5% of the width) of chamfering and recession are very effective in suppressing the aeroelastic instabilities whereas larger modifications have an adverse effect and promote the instabilities at lower velocities. They concluded that the effectiveness of small aerodynamic modifications is associated with the enhanced aerodynamic damping and is not due to suppression of the vortex shedding phenomenon.

Tamura et al. (1998) conducted the numerical and experimental investigation over the square cylinder with aerodynamic modifications of corner chamfering (16.67%) and corner roundness (16.67%). Through a detailed investigation, they demonstrated that the essential aerodynamic characteristics (like an average drag, fluctuating lift forces, and Strouhal number) and their statistical distribution could be captured with the help of 3D numerical computations. However, for the unsteady characteristics like peak pressure and the especially the span-wise correlations of these unsteady pressure are very difficult to measure. Most of the researchers claim the reason for this as insufficient discretization in span-wise direction, but it is still a chance for various other possibilities which may cause this discrepancy. Tamura et al. (1998) tried to perform a high computational study with a higher order of discretization to check the truthfulness for the above claim. They found that averaged pressure and fluctuating pressure on the sharp-edged and modified cylinders can provide sufficient accuracy for stimulations which adopt very high grid resolutions. Effect of both types of modification on the mean pressure is to reduce the pressure values drastically near the corners even on the windward side whereas the increase in RMS pressures values can also be observed in the same region. In case of the span-wise correlations for sharp corners, they found that even with very high grid resolution it is difficult to capture the exact trend of a sudden decrease in the correlation with the increase in the span-wise length as found in the experiments. They suggested that these variations may be due to the difference in the turbulence levels and the way they are accommodated in the numerical simulations. In case of modified cylinders, they found that the span-wise correlations are reduced for chamfered cylinder whereas it is increased for cornered rounding modification.

Tamura and Miyagi (1999) performed an experimental investigation to ascertain the effect of corner chamfering (16.67%) and corner roundness (16.67%) over a square cylinder in uniform flow conditions. Aerodynamic forces were measured at the base of the model with the help of 3-component load cells, and they concluded that (a) Chamfering reduces the Coefficient of drag and rounding of corners results in further reduction of the drag coefficient. (b) Orientation at which the slope of lift coefficient changes its sign (i.e., approximately the orientation at which reattachment occurs) for the square section is found to be around 12° which is similar as reported by previous researchers. Chamfering brings down this orientation to 7°–8° and rounding further brings it down to 5° orientation, which means aerodynamic modifications promote the reattachments at lower orientations. (c) Aerodynamic modifications significantly reduce the fluctuation lift component for the lower range of orientations.

Choi and Kwon (2001) performed a detailed investigation over a rectangular cylinder having side ratio (B/A = 1.0, 1.5 and 2.0) with the aerodynamic modification of single recession with proportions varying from 4% to 20% of the depth of the section. Wind tunnel test was performed in uniform flow for orientations varying from 0° to 45° at an interval of 5°, the vortex induced responses, and Strouhal Number was measured. They concluded that, in cases of recession with 4-10%, Strouhal number for the attack angles from 15° to 45° have the smaller Strouhal number values than the original section with no corner cut, which implies that these proportions of modification are ineffective to stabilize the vortex-induced vibration. They claimed that a particular modification is effective if the flow which gets separated at the leading edge of recession most probably gets reattached near the second edge of recession minimizing the separation width in wake which further leads to increased Strouhal Number. Based on this they were able to identify that for orientation of 20° and 25°, 12% to14% and 16% to 20% of modification respectively were effective as they resulted into higher Strouhal number than original section. Thus, for the section with (B/A = 1.0) with single recession, they were able to classify the orientations from 15° to 45° into two regions according to the effectiveness of modification as namely; (a) Effective region (15°–25° Region-I) and (b) Ineffective region (30°–45° Region-II).

Kim and You (2002) investigated the effect of tapering along with the height of building with 5%, 10% and 15% of modification in the urban and suburban flow conditions. Effect of tapering is more in across wind direction as compared to the along-wind response of the building. An important observation was that the tapering does not necessarily reduce the responses when compared to the square building and the effectiveness is highly dependent of the orientation of the building (angle of attack of wind) and proportioning adopted.

A numerical study was conducted by Dalton and Zheng (2003) over a square cylinder with aerodynamic modification of corner roundness (D/8) for two orientations of 0° and 45°. Effect of modification over vortex shedding pattern and the aerodynamic forces were evaluated in the uniform flow conditions. They found a significant reduction in the drag, which was even more in the case of 45° orientation. In the case of corner rounded section, they observed that the rolling up of the shear layer was not as intense as in sharp-edged square section and hence the base suction was relatively weak reducing the drag coefficient.

Gu and Quan (2004) performed an experimental investigation with high-frequency force balance technique over 15 building models of 300 height. Effect of aspect ratio (H/B = 4 to 9), side ratio (0.33, 0.5, 0.67, 1.50, 2.0, 3.0) and aerodynamic modification like chamfering and single recession with cut proportion of (5%,10%and 20% of width) over the first mode of across wind aerodynamic forces were evaluated. From a comparison of non-dimensional power spectra for a crosswind, it was concluded that (a) increase in aspect ratio results into higher frequency and amplitude of the spectral peak. (b) for side ratio less than unity the peak amplitude of power spectra for crosswind forces is higher with narrow band whereas peak amplitudes are lowered with a broader band for side ratio higher than one. (c) The effect of both types of modification is very significant and helps to reduce the peak amplitudes of the power spectra for crosswind forces and 10% modification performing the best of the compared proportions.

Kim et al. (2008) extended the study of the tapered model into the aero-elastic regime and tested three models of similar tapering proportions of 5%, 10% and 15% like Kim and Young (2003). They stated that the effect of tapering could be noticed for the building with moderate damping (2%–4%) when the reduced velocities are on the comparatively higher side. They confirmed that in certain situations like for buildings with lower damping, the effect of an increase in tapering leads to adverse effects like increased across wind fluctuating displacement responses.

Yamagishi et al. (2009) presented results of a numerical and experimental study performed to understand the characteristics of flow over building with five corner modifications namely; corner recession 10% of the width, corner arc 10% of the width and corner chamfering 3.33%, 10%, and 16.67% of the width. A numerical study was conducted with FLUENT 6.3 with RNG K-Ɛ model adopted to model turbulence, whereas the visualization study was made with oil film and mist flow method. Yamagishi et al. (2009) concluded that there is a significant effect of angle of attack on drag coefficient and corner cut (chamfering by 10% of the width) reduces the drag by about 30% than that of a square cylinder. Though the pressure distribution around the cylinders was plotted, the effect of the modifications on it was not discussed.

Kurata et al. (2009) studied the effect of recession and side ratio of the rectangular cylinder on the drag coefficient. The breadth (B) of the cylinder is kept constant, and the length (A) was varied to obtain (B/A) ratios of (0.083 to 1.0) with an equal increment of (0.083, i.e., 1/12). The corner cuts are provided on front edge only with the dimensions denoted as C1 and C2 in the Y and X directions respectively. In the experiment, C1/A was varied from 0 to 0.20, and C2/A was varied from 0 to 0.25 with an increment of 0.05, finally resulting in 21 set of models. For the models without any recession, they confirmed the observations of Bearman and Trueman (1972) stating the drag attains its maximum value at the side-ratio of 0.68 for the zero-degree angle of attack and wider side facing the wind flow. In the case of all combinations of single recession models with a zero-degree angle of attack, the results showed that the Cd value plotted against C1/A gives the convex downward profile. Also, the minimum value of Cd and hence, the drag force will lie in the selected range of C1/A between 0 to 0.2 for all the tested cases of side ratio (B/A).

Ueda et al. (2009) studied the physical mechanism behind the drag reduction with the help of single recession through visualization techniques such as; experimental (aluminum flake technique, hydrogen bubble technique, and oil flow technique) and numerical (2-D vortex particle method).Three kinds of recession configurations ((i) C1/A = C2/A = 0.0 (ii) C1/A = C2/A = 0.25 and (iii) C1/A = 0.10 and C2/A = 0.20), were adopted to visualize the phenomenon of separation and reattachment of shear layer from front corner. Through this study, Ueda et al. (2009) conclude that there is some backflow on the side wall in case (i) when no corner cuts are provided resulting into the widening of wake width as shown in the Fig. 11. In case (ii) the flow gets separated at first corner 'C' and tries to reattach on the side wall, but again gets separated at corner 'A' and result into almost similar kind of backflow on the side wall and wide wake region. Whereas in case (iii) the flow after separation at corner 'C' manages to reattach on the side wall and flow along the side wall without any backflow, reducing the wake width and hence effective in reducing the drag. Finally, the author has proposed an equation for optimum corner cut based on free streamline theory, with the assumption that an optimal drag reduction is possible by making sure that the separation streamline reattaches the sidewall and flow along with it.

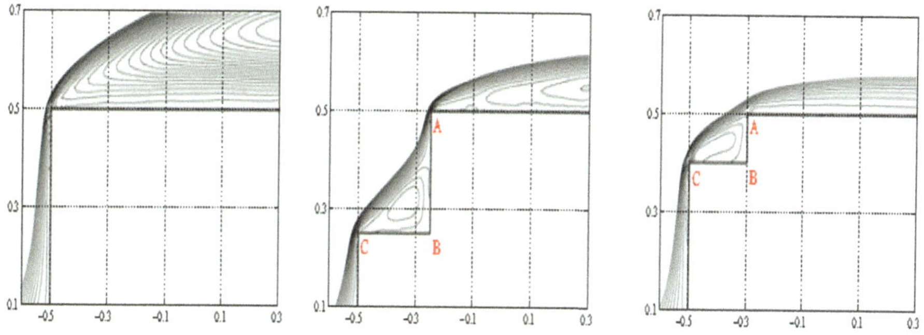

Fig. 11. Effect of corner recession on the backflow over the side face of the square cylinder (Source: Ueda et al. (2009))

Ueda et al. (2012) continued the experimental study carried out by Kurata et al. (2009) and Ueda et al. (2009) and focuses on the effect of corner cuts on drag coefficients when the angle of attack is not limited to perpendicular to the face of the building but varies from 0° to 45°. The difference is also that the corner cuts are provided on all the sides and not only on the front face. Three corner recession configurations sizing (i) C1/A = C2/A = 0.0 (ii) C1/A = 0.15 and C2/A = 0.05 and (iii) C1/A = 0.15 and C2/A = 0.25 were studied. Experimental results showed that case (iii) configuration was able to achieve the effective drag reduction when the angle of attack was in between 0° to the angle of 14° whereas the case (ii) showed results very similar to that of the square prism and did not cause any notable reduction in drag.

Through flow visualization study it was observed that for case (i) and case (ii) the separated shear layer caused a strong backflow on side wall resulting into wide wake whereas in case (iii) the separated shear layer reattached on the sidewall resulting into narrow wake width and hence the drag was reduced. Authors from the above observation concluded that the shear layer generated at the front corner play an important role in drag reduction even when the angle of attack varies.

Carassale et al. (2012), Carassale et al. (2014) performed experimental investigation over a square cylinder and two aerodynamically modified cylinders (corner rounding of 6.6% and 13.33%) and both global effects like drag, lift, fluctuating lift and Strouhal Number and localized surface pressure distribution were measured for various orientations varying from 0° to 45°. They observed that for sharp-edged cylinder two characteristic flow regimes (completely separate flow and reattached flow regimes) and are identified and separated by a critical angle of incidence, α_{cr} located about 12–15°. Increase in corner roundness promotes the reattachment of the flow on the side face resulting in a reduction of the critical angle of incidence associated with the reattachment flow. This results in the reduction of drag, lift and fluctuating lift coefficients and increase in Strouhal Number as shown in Fig. 12. Measurement of the mean surface pressure showed that in case of turbulent flow and subcritical regime (Re = 2.49 × 104 and 5.10 × 104) the Cp variations are is very similar to Cp obtained in smooth flow the same analogy also applies to the fluctuating pressure coefficients.

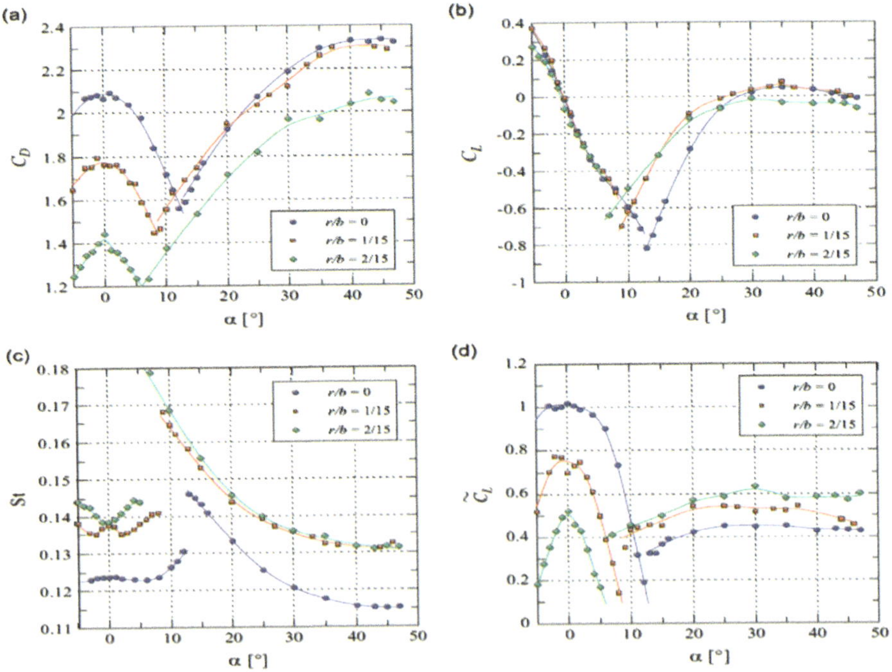

Fig. 12. Effect of corner roundness on (a) Drag, (b) Lift, (c) Strouhal number, and (d) Fluctuating lift (Source: Carassale et al. (2012))

Tanaka et al. 2012 performed an extensive experimental investigation on various models building model with a variety of basic plan shape, with and without aerodynamic modifications in the horizontal and vertical plane, models tilted along with the height and so on. They studied the effect of these variations on both global effects like mean overturning moments, fluctuating overturning moments, Strouhal number; and localized effects like mean pressure distribution. Some important conclusions of the study can be listed as (a) Tapped models such as 4-side tapered and tapered with setbacks are effective in reducing the mean overturning moment in along-wind direction whereas the corner modified models, and helical models are effective in reducing across wind overturning moments. (b) The vortex shedding for square base model occurs at almost the similar frequency at all the heights whereas, in case of tapered and helical models, it varies significantly along with the height. As a result, these modifications are very effective in tackling the issues related to vortex-induced vibrations. (c) In case of corner recessed models, the largest negative mean pressure occurs near the leading edge on the side face and is significantly different from that of the base model whereas the mean pressure on the rear face shows reduced suction on the rear face than the base model as shown in the Fig. 13.

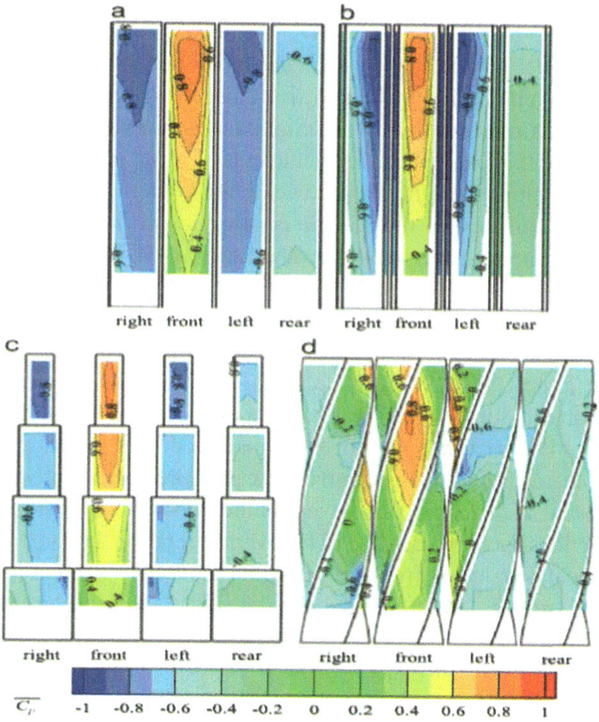

Fig. 13. Distribution of mean wind pressure coefficients (a) Square, (b) Recession, (c) Setback and (d) 1801 Helical (Source: Tanaka et al. (2012))

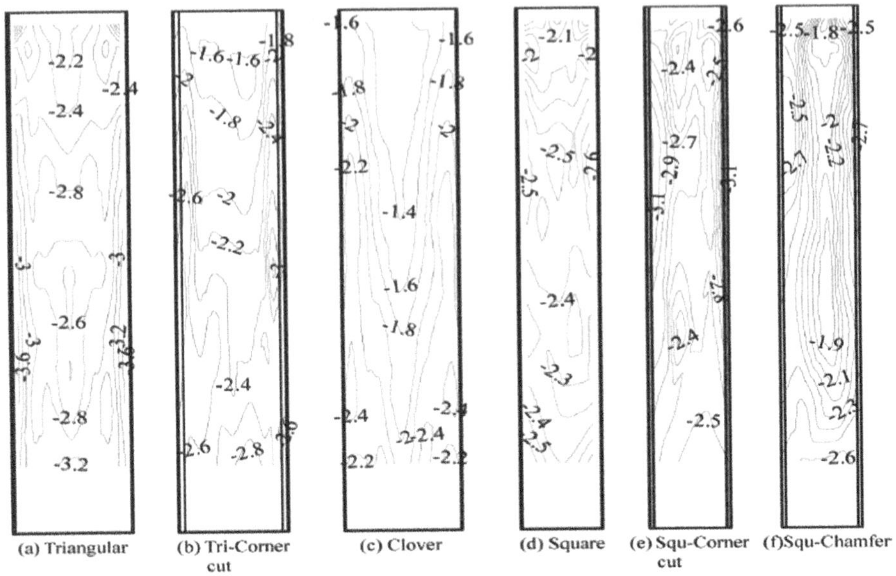

Fig. 14. Comparison of peak negative pressure coefficients for various modifications (Source: Bandi et al. (2013))

Bandi et al. (2013) performed a detailed experimental investigation on various building models by classifying them into five categories: polygonal, tapered, helical, corner modified, and combinations of all. Significant observations of the study can be listed as (a) The positive peak pressure coefficients occur at upper levels for all models and its distribution varies smoothly from lower to the higher values from corner regions to the center of the surfaces for the triangle-corner cut and clover models whereas it is almost constant for the square models. (b) The largest peak suctions occur close to the corner regions for all the models, and it occurs between 0.4H–0.78H (where 'H' is the height of the building) for the triangle-corner cut, clover, square, square-corner cut (recession) and square-chamfered models, whereas for the triangular model it occurs at 0.98H. (c) The largest negative peak pressure coefficient for the triangle-corner cut model is less than that for the triangular base model, whereas for the square-corner cut (recession) model is higher than that for the square model, as shown in Fig. 15. (d) For helical models, the distribution of negative peak pressure coefficient varies widely, and peak suctions occur at the corners and even on the mid-surfaces, but the area occupied by the peak suctions is very small. (e) The overall trend of the largest negative peak pressure coefficient is reducing from the triangular model to a circular model, i.e., when the number of surfaces increases the largest negative peak pressure coefficient decreases (Fig. 14).

Kim and Kanda (2013) investigated the effect of tapering modification over spatiotemporal variations of fluctuating pressure. Four scaled building models (square, square with a setback at mid-height and two tapered models (5% and 10% of tapering ratio) were tested in the wind tunnel for orientation of 0° to 180° with an interval of

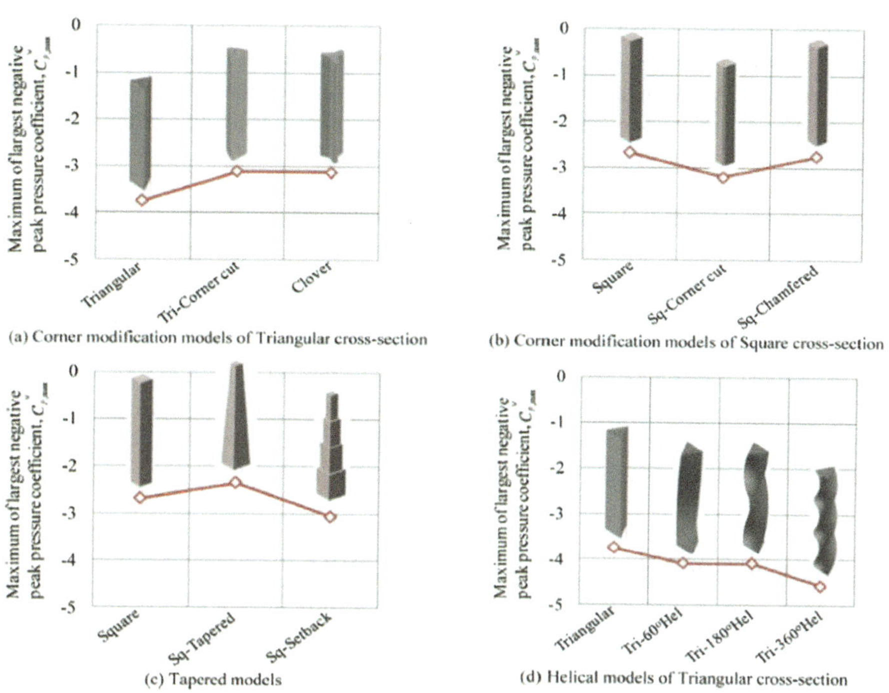

Fig. 15. Comparison of largest peak negative pressure coefficients for various modifications (Source: Bandi et al. (2013))

15°. There is almost no effect on the mean pressure distribution over the windward face in all the compared configurations whereas some difference can be seen over the leeward face in the upper region there is where a little increase in the suction was observed in the modified configurations. Adverse impact (increased suction) in the model with a setback in the region near modification was observed, which was quantified as peak negative pressure coefficient of −4.3 in place of −3.8 for the square building model.

He et al. (2014) executed a flow visualization studies using particle image velocimetry to understand the mechanism for drag reduction on the corner recessed square cylinder. From detailed analysis they found that wake width is reduced, the length of the recirculation region is increased, and the fluctuations in velocity in the wake region are weakened as compared to the square cylinder without a recession. The drag coefficient (Cd) for corner recessed configuration was reduced in the range of about 13 to 44%, and Cd was showed to be linearly proportional to the minimum wake width. He et al. (2014) also confirmed that corner recession persuades the vortices formation to take place further on downstream than that of the square cylinder without corner recession. They explained the physical mechanism of flow and reveal that the pair of recirculation flow formed in the corner cut area act as virtual shape and helps the flow pass the corners without much separation and thus drag is reduced.

Miran and Sohn (2015) conducted a numerical study over a square cylinder with the aerodynamic modification of corner rounding (R/D = 0.1 to 0.5). Aerodynamic characteristics like Strouhal number, drag coefficient, and fluctuating lift coefficient were evaluated and compared with existing literature. Mean drag coefficient and fluctuating lift coefficient are found to be minimum for (R/D = 0.2). Physical reasoning for this was obtained by observing the instantaneous stream-wise vortices contours, which indicates that the alternate vortices shed for the modified cylinders (R/D > 0.2) do not get deflected away from the centerline of the wake resulting into smaller wake widths and hence lowered drag on the cylinder.

Miran and Sohn (2016) continued numerical study over the square cylinder with corner rounding to evaluate the effect of inclination of the cylinder. They confirmed that the critical orientation, which promotes the reattachment of flow and hence reduces the drag coefficient, reduces fluctuating lift coefficient, and increases the Strouhal number would occur at 12° as established in the literature. Rounding of corners brings down this critical orientation further down in a range of 5° to 10°.

4 Critical Analysis of Reviewed Literature

After providing a detailed overview of most of the relevant papers in the subject domain, this section tries to evaluate the effective progress achieved in particular area of wind effects on building with and without aerodynamic modification especially from the perspective of façade designer. The Fig. 16 shows the decade wise distribution of the reviewed papers; it traces the direction of evolution that has taken place in the specified area of research. It gives a picture of how initially in the 1970s and 1980s a large amount of research work was performed on wind evaluation of buildings with basic shapes. Moreover, later in the 1990s and early 20th century, the research trend shifted towards the evaluation of building with modifications. However, when Fig. 16 is seen along with Fig. 17, which segregates the reviewed research articles based on whether the research contributed in the global or localized evaluation of wind effect, then the picture is not very encouraging for façade designer. It clearly shows that though a considerable amount of research is being done in the area of wind evaluation on building with aerodynamic modifications, only a few articles actually try to address the issues about localized effects. Following are typical limitations identified in most of the research works reviewed:

(1) Limitation-1: Most of the research works, especially the one with aerodynamic modifications, are targeted on the evaluation of the global aspects such as along-wind and across-wind overturning moment, power spectral densities and trajectories of various wind force coefficients. These studies though have proved very useful for enhancing the safety and economic aspects in the designing of the tall buildings as a whole but from the perspective of façade designer very little quantitative information is available on the localized effects like temporal and spatial distribution of surface pressures.

(2) Limitation-2: Even in most of the articles where the surface pressure distribution was measured, the discussions were based on the mean pressure distribution which

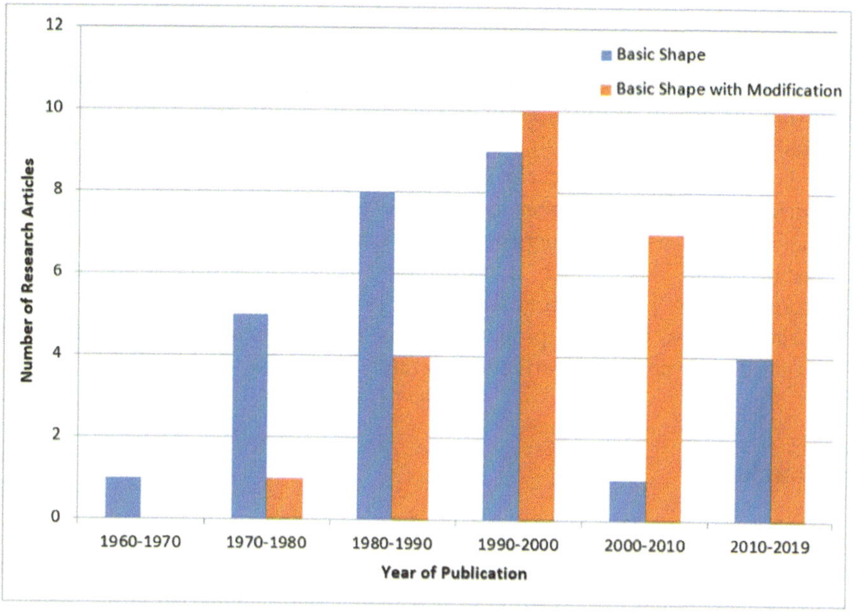

Fig. 16. Decade-wise distribution of the reviewed research articles

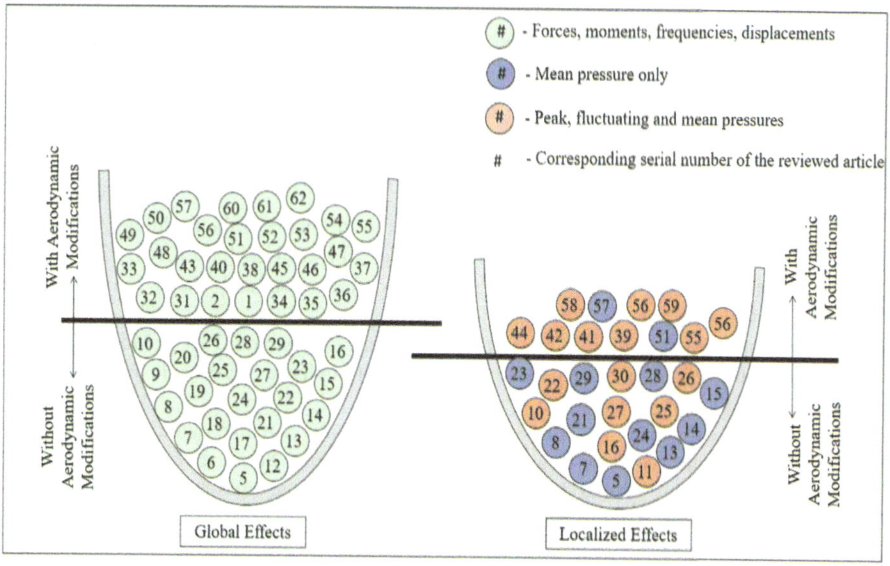

Fig. 17. Segregation of the reviewed research articles based on the types of wind effects considered for evaluation

was further used for better understanding of the flow characteristics and the global phenomenon like vortex shedding frequency and induced vibrations. The intricate pressure characteristics like fluctuating pressure or peak pressure values, which will govern the design of glass panels in cladding or façade were not focused.

(3) Limitation-3: One of the significant limitations of the articles which performed studies to evaluate the effect of aerodynamic modification was that these studies generally used one member from each geometric modification family (like chamfering, recession, corner roundness, or tapering) and then compared it with a member from other. Such investigations, although provide some qualitative information, are insufficient to address the practical issue of optimization faced by façade designer. Effect of proportioning of the modification, especially over the localized peak pressure distribution, is rarely addressed.

5 Conclusion

The present paper provides a comprehensive overview of all the types of studies performed like wind tunnel experimental evaluation (quantitative) flow visualization study (qualitative) and numerical (both quantitative and qualitative) in the domain of wind evaluation of tall buildings with and without aerodynamic modifications. The critical analysis traces the direction of evolution of the research in this specific area of study and ascertains the potential research gaps by identifying the typical limitations in the existing literature. Paper establishes a need for benchmark studies to be performed, which has a specific focus on the localized effects like peak pressure estimation for building configurations having aerodynamic modification with various proportions thereby ensuring safe and economical design of building façades.

References

Amin, J.A., Ahuja, A.K.: Aerodynamic modifications to the shape of the buildings: a review of the state-of-the-art. Asian J. Civ. Eng. (Build. Hous.) **11**(4), 433–450 (2010)

Mooneghi, M.A., Kargarmoakhar, R.: Aerodynamic mitigation and shape optimization of buildings. J. Build. Eng. **6**, 225–235 (2016)

Basu, R.I.: Aerodynamic forces on structures of circular cross-section. Part 1. Model-scale data obtained under two-dimensional conditions in low-turbulence streams. J. Wind Eng. Ind. Aerodyn. **21**(3), 273–294 (1985)

Demartino, C., Ricciardelli, F.: Aerodynamics of nominally circular cylinders: a review of experimental results for Civil Engineering applications. Eng. Struct. **137**, 76–114 (2017)

Vickery, B.J.: Fluctuating lift and drag on a long cylinder of square cross-section in a smooth and in a turbulent stream. J. Fluid Mech. **25**(3), 481–494 (1966)

Hunt, J.C.R.: The effect of single buildings and structures. Philos. Trans. R. Soc. London Ser. A Math. Phys. Sci. **269**(1199), 457–467 (1971)

Lee, B.E.: The effect of turbulence on the surface pressure field of a square prism. J. Fluid Mech. **69**(2), 263–282 (1975)

Huot, J.P., Rey, C., Arbey, H.: Experimental analysis of the pressure field induced on a square cylinder by a turbulent flow. J. Fluid Mech. **162**, 283–298 (1986)

Nakamura, Y., Ohya, Y.: The effects of turbulence on the mean flow past two-dimensional rectangular cylinders. J. Fluid Mech. **149**, 255–273 (1984)

Kareem, A., Cermak, J.E.: Pressure fluctuations on a square building model in boundary-layer flows. J. Wind Eng. Ind. Aerodyn. **16**(1), 17–41 (1984)

Kareem, A.: Measurements of pressure and force fields on building models in simulated atmospheric flows. J. Wind Eng. Ind. Aerodyn. **36**, 589–599 (1990)

Okajima, A.: Strouhal numbers of rectangular cylinders. J. Fluid Mech. **123**, 379–398 (1982)

Chen, J.M., Liu, C.H.: Vortex shedding and surface pressures on a square cylinder at incidence to a uniform air stream. Int. J. Heat Fluid Flow **20**(6), 592–597 (1999)

Norberg, C.: Flow around rectangular cylinders: pressure forces and wake frequencies. J. Wind Eng. Ind. Aerodyn. **49**(1–3), 187–196 (1993)

Steggel, N.: A numerical investigation of the flow around rectangular cylinders. Doctoral dissertation, University of Surrey (1998)

Igarashi, T.: Characteristics of the flow around rectangular cylinders: the case of the angle of attack 0 deg. Bull. JSME **28**(242), 1690–1696 (1985)

Davis, R.W., Moore, E.F.: A numerical study of vortex shedding from rectangles. J. Fluid Mech. **116**, 475–506 (1982)

Franke, R., Rodi, W., Schönung, B.: Numerical calculation of laminar vortex-shedding flow past cylinders. J. Wind Eng. Ind. Aerodyn. **35**, 237–257 (1990)

Okajima, A.: Numerical simulation of flow around rectangular cylinders. J. Wind Eng. Ind. Aerodyn. **33**(1–2), 171–180 (1990)

Ohya, Y., Nakamura, Y., Ozono, S., Tsuruta, H., Nakayama, R.: A numerical study of vortex shedding from flat plates with square leading and trailing edges. J. Fluid Mech. **236**, 445–460 (1992)

Bearman, P.W., Trueman, D.M.: An investigation of the flow around rectangular cylinders. Aeronaut. Q. **23**(3), 229–237 (1972)

Akins, R.E., Cermak, J.E.: Wind pressures on buildings. CER; 76/77-15 (1976)

Laneville, A., Yong, L.: Mean flow patterns around two-dimensional rectangular cylinders and their interpretation. J. Wind Eng. Ind. Aerodyn. **14**(1–3), 387–398 (1983)

Ohya, Y.: Note on a discontinuous change in wake pattern for a rectangular cylinder. J. Fluids Struct. **8**(3), 325–330 (1994)

Bruno, L., Fransos, D., Coste, N., Bosco, A.: 3D flow around a rectangular cylinder: a computational study. J. Wind Eng. Ind. Aerodyn. **98**(6–7), 263–276 (2010)

Robertson, J.M., Wedding, J.B., Peterka, J.A., Cermak, J.E.: Wall pressures of separation—reattachment flow on a square prism in uniform flow. J. Wind Eng. Ind. Aerodyn. **2**(4), 345–359 (1978)

Igarashi, T.: Characteristics of the flow around a square prism. Bull. JSME **27**(231), 1858–1865 (1984)

Knisely, C.W.: Strouhal numbers of rectangular cylinders at incidence: a review and new data. J. Fluids Struct. **4**(4), 371–393 (1990)

Dutta, S., Muralidhar, K., Panigrahi, P.: Influence of the orientation of a square cylinder on the wake properties. Exp. Fluids **34**(1), 16–23 (2003)

Lipecki, T., Bęc, J., Błazik-Borowa, E.: Surface pressures on rectangular cylinders–the dependence on aspect ratio, wind structure, and angle of wind attack. In: The Seventh International Colloquium on Bluff Body Aerodynamics and Applications (BBAA7) Shanghai, China, 2–6 September 2012 (2012)

Davenport, A.G.: The response of six building shapes to turbulent wind. Philos. Trans. R. Soc. London Ser. A Math. Phys. Sci. **269**(1199), 385–394 (1971)

Kareem, A.: Mitigation of wind induced motion of tall buildings. J. Wind Eng. Ind. Aerodyn. **11**(1–3), 273–284 (1983)

Kwok, K.C., Bailey, P.A.: Aerodynamic devices for tall buildings and structures. J. Eng. Mech. **113**(3), 349–365 (1987)
Kwok, K.C.S.: Effect of building shape on wind-induced response of tall building. J. Wind Eng. Ind. Aerodyn. **28**(1–3), 381–390 (1988)
Kwok, K.C.S., Wilhelm, P.A., Wilkie, B.G.: Effect of edge configuration on wind-induced response of tall buildings. Eng. Struct. **10**(2), 135–140 (1988)
Hayashida, H., Iwasa, Y.: Aerodynamic shape effects of tall building for vortex induced vibration. J. Wind Eng. Ind. Aerodyn. **33**(1–2), 237–242 (1990)
Dutton, R., Isyumov, N.: Reduction of tall building motion by aerodynamic treatments. J. Wind Eng. Ind. Aerodyn. **36**, 739–747 (1990)
Okamoto, S., Uemura, N.: Effect of rounding side-corners on aerodynamic forces and turbulent wake of a cube placed on a ground plane. Exp. Fluids **11**(1), 58–64 (1991)
Jamieson, N.J., Carpenter, P., Cenek, P.D.: Wind induced external pressures on a tall building with various corner configurations. J. Wind Eng. Ind. Aerodyn. **44**(1–3), 2401–2412 (1992)
Miyashita, K., Katagiri, J., Nakamura, O., Ohkuma, T., Tamura, Y., Itoh, M., Mimachi, T.: Wind-induced response of high-rise buildings effects of corner cuts or openings in square buildings. J. Wind Eng. Ind. Aerodyn. **50**, 319–328 (1993)
Luo, S.C., Yazdani, M.G., Chew, Y.T., Lee, T.S.: Effects of incidence and afterbody shape on flow past bluff cylinders. J. Wind Eng. Ind. Aerodyn. **53**(3), 375–399 (1994)
Surry, D., Djakovich, D.: Fluctuating pressures on models of tall buildings. J. Wind Eng. Ind. Aerodyn. **58**(1–2), 81–112 (1995)
Kawai, H.: Effect of corner modifications on aeroelastic instabilities of tall buildings. J. Wind Eng. Ind. Aerodyn. **74**, 719–729 (1998)
Tamura, T., Miyagi, T., Kitagishi, T.: Numerical prediction of unsteady pressures on a square cylinder with various corner shapes. J. Wind Eng. Ind. Aerodyn. **74**, 531–542 (1998)
Tamura, T., Miyagi, T.: The effect of turbulence on aerodynamic forces on a square cylinder with various corner shapes. J. Wind Eng. Ind. Aerodyn. **83**(1–3), 135–145 (1999)
Choi, C.K., Kwon, D.K.: The characteristics of Strouhal number of rectangular cylinders with various corner cuts. JWE **89**, 153–156 (2001)
Kim, Y.M., You, K.P.: Dynamic responses of a tapered tall building to wind loads. J. Wind Eng. Ind. Aerodyn. **90**(12–15), 1771–1782 (2002)
Dalton, C., Zheng, W.: Numerical solutions of a viscous uniform approach flow past square and diamond cylinders. J. Fluids Struct. **18**(3–4), 455–465 (2003)
Gu, M., Quan, Y.: Across-wind loads of typical tall buildings. J. Wind Eng. Ind. Aerodyn. **92**(13), 1147–1165 (2004)
Kim, Y., You, K., Ko, N.: Across-wind responses of an aeroelastic tapered tall building. J. Wind Eng. Ind. Aerodyn. **96**(8–9), 1307–1319 (2008)
Yamagishi, Y., Kimura, S., Oki, M., Hatayama, C.: Effect of corner cutoffs on flow characteristics around a square cylinder. In: 10th International Conference on Fluid Control, Measurements, and Visualization (FLUCOME), Moscow, Russia (2009)
Kurata, M., Ueda, Y., Kida, T., Iguchi, M.: Drag reduction due to cut-corners at the front-edge of a rectangular cylinder with the length-to-breadth ratio being less than or equal to unity. J. Fluids Eng. **131**(6), 064501 (2009)
Ueda, Y., Kurata, M., Kida, T., Iguchi, M.: Visualization of flow past a square prism with cut-corners at the front-edge. J. Vis. **12**(4), 383–391 (2009)
Ueda, Y., Kurata, M., Kida, T., Iguchi, M.: Flow past a square prism with cut-corners (effect of the angle of attack). J. Japan. Soc. Exp. Mech. **12**, s23–s28 (2012)
Carassale, L., Freda, A., Brunenghi, M.M., Piccardo, G., Solari, G.: Experimental investigation on the aerodynamic behavior of square cylinders with rounded corners. In: The Seventh

International Colloquium on Bluff Body Aerodynamics and Applications (BBAA7) Shanghai, China; 2–6 September 2012 (2012)
Carassale, L., Freda, A., Marre-Brunenghi, M.: Experimental investigation on the aerodynamic behavior of square cylinders with rounded corners. J. Fluids Struct. **44**, 195–204 (2014)
Tanaka, H., Tamura, Y., Ohtake, K., Nakai, M., Kim, Y.C.: Experimental investigation of aerodynamic forces and wind pressures acting on tall buildings with various unconventional configurations. J. Wind Eng. Ind. Aerodyn. **107**, 179–191 (2012)
Bandi, E.K., Tanaka, H., Kim, Y.C., Ohtake, K., Yoshida, A., Tamura, Y.: Peak pressures acting on tall buildings with various configurations. Int. J. High-Rise Build. **2**(3), 229–244 (2013)
Kim, Y.C., Kanda, J.: Wind pressures on tapered and set-back tall buildings. J. Fluids Struct. **39**, 306–321 (2013)
He, G.S., Li, N., Wang, J.J.: Drag reduction of square cylinders with cut-corners at the front edges. Exp. Fluids **55**(6), 1745 (2014)
Miran, S., Sohn, C.H.: Numerical study of the rounded corners effect on flow past a square cylinder. Int. J. Numer. Methods Heat Fluid Flow **25**(4), 686–702 (2015)
Miran, S., Sohn, C.H.: Influence of incidence angle on the aerodynamic characteristics of square cylinders with rounded corners: a numerical investigation. Int. J. Numer. Methods Heat Fluid Flow **26**(1), 269–283 (2016)

Experimental Investigation of Two-Way Hinges in Reinforced Concrete Members

Safwan Chahal[1](✉), Oussama Baalbaki[1], Yehya Timsah[1], Hassan Ghanem[1], and Zaher Abu Saleh[2]

[1] Beirut Arab University, Beirut, Lebanon
safwanchahal79@gmail.com
[2] Rafic Hariri University, Beirut, Lebanon

Abstract. Concrete foundations in seismic regions are usually designed to withstand plastic hinge moments that can be developed at the bases of the columns. These hinges are used to eliminate the moments transferred to the foundations. The ratio adopted by the American Concrete Institute (ACI) to determine the amplification of the strength of the hinge due to confinement effects, is limited to two. This paper presents the experimental analysis conducted on several two-way hinges to assess their behavior on concrete structures with specific focus on the amplification factor. The main parameters used in this investigation are the square root of the column area over the hinge area, and the overall hinge strength due to vertical throat concrete crushing failure mode. It is expected that the results of this study will lead to the development of guidelines for the behavior of two-way hinges in reinforced concrete structures.

Keywords: Reinforced concrete · Hinges · Amplification factor · Compressive stress-strain relationship

1 Introduction

1.1 Introduction

Structural concrete hinges are used to prevent the transfer of bending moments. This is intended to either reduce stresses transferred to the foundations in steel or bridge structures or to redistribute the forces and stresses applied to the structures.

Hinges are formed by several means. A reduction in the concrete cross-section is usually sufficient to produce hinge like behavior. The stresses in the smaller section due to bending moments reach ultimate capacity quickly, thus causing cracks and loss of sustaining further bending moment (i.e. hinge like behavior). The primary benefit of hinges is the reduction of moments on the foundation, thus reducing the cost of the foundation. An additional benefit is the lower cost of repairing a damaged two-way hinge compared to that of a damaged full moment connection [1, 2].

Development in reinforced concrete hinges was significant the past decades. However, there are many ambiguities, undiscovered items and misunderstood aspects. Many design considerations or aspects are either conservative, empirical or avoided. Two way hinges are also not considered extensively in literature and codes.

H. Rodrigues et al. (Eds.): GeoMEast 2019, SUCI, pp. 128–146, 2020.
https://doi.org/10.1007/978-3-030-34216-6_9

This paper presents the experimental analysis conducted on several two-way hinges to assess their behavior on concrete structures with specific focus on the amplification factor. The main parameters used in this investigation are the square root of the column area over the hinge area, and the overall hinge strength due to vertical throat concrete crushing failure mode. As the main purpose is only to compare the confinement effects, the bending and splitting failure of the column were prevented. This gives a better understanding of the confinement effect on the hinged area.

1.2 Literature Review

October 1990, David I. McLean and Kuang Y. Lim experimentally investigated the effects on two-way hinges performance of several parameters, including vertical discontinuity in the hinge detail, level of axial load, low-cycle fatigue characteristics, column aspect ratio, and different amounts of longitudinal and transverse reinforcement. They showed that hinge details can be incorporated into columns to significantly reduce the moment capacity at the bases of the columns. However, the moments are not negligible, as is sometimes assumed for design with the moment-reducing hinge details. Providing vertical discontinuity in the hinge resulted in reduced distress in the longitudinal reinforcement and improved the performance of the hinge for design with the moment-reducing hinge details. Providing vertical discontinuity in the hinge resulted in reduced distress in the longitudinal reinforcement and improved the performance of the hinge [3].

June 2006, Zhiyuan Cheng, M. Saiid Saiidi, and David H. Sanders, experimentally investigated the performance of circular hinges subjected to combine vertical and lateral loads including seismic forces, and developed a design method for practical application. They showed that regardless of the level of axial load, size of hinge, hinge steel ratio, and the column aspect ratio, the shear capacity of two-way hinges is much lower than that the shear friction theory estimates [1, 4].

December 2010, Arash E. Zaghi and M. Saiid Saiidi, investigated the Seismic Performance of Pipe-Pin Two-Way Hinges in Concrete Bridge Columns. They proposed a rational method based on the possible limit states to obtain the lateral capacity of these hinges. They observed that pipe-pin two-way hinges perform as moment-free connections while transferring shear and axial loads. The experimental results confirmed that the hinges designed based on the proposed guideline remain elastic with no damage. The pipes were straight, intact, and with no sign of damage even after drift ratios exceeding 8% that led to the failure of the test model at column plastic hinges [5–7].

July 2014, A. Mehrsosouch and M. S. Saiidi investigated the cyclic behavior and performance of a pipe pin model incorporated in a two-column pier model. The results showed that pipe pins remained essentially elastic and could be considered as capacity protected members. Moreover, pipes pins performed well in transferring the shear forces to the footing through interaction of the pipes and friction. Closer to the hinge led to higher plastic shear in the column and the pins. The damage was limited to spalling of concrete at edge of the columns, which was repairable [8, 9].

April 2015, Gregor Schacht and Steffen Marx described the historical development of one-way concrete hinges and the experimental investigations that have been carried out over the past century. They explained the main differences in concepts and the load-bearing behaviours between Freyssinet and Mesnager one-way concrete hinges. They showed that these concrete hinges are suitable for many different bridge types [10].

1.3 Goals and Objectives, Novelty of the Research

The main purpose of this article is (1) to experimentally investigate the behavior of the two-way hinges subjected to axial load, (2) to determine the confinement effects imposed by the larger column area on the smaller two-way hinge throat area, and (3) to propose a reliable design method for practical application in reinforced concrete column-hinge connection.

2 Experimental Program

2.1 Introduction

An experimental study was conducted on eighteen 1/3-scale specimens which were subjected to a uniform axial load only. The load-displacement curves of the specimens are then plotted online allowing the monitoring of the behavior. These specimens represent several confinement levels, which are used to determine the relation between the square root of the column area (A_c) over the hinge area (A_h) and the overall hinge strength due to vertical throat concrete crushing failure.

2.2 Test Specimens

The specimens are divided into three major series (CS-22, CS-32, and CS-40). Each series consists of six two-way reinforced concrete hinges specimens. The geometry, reinforcement distribution and mechanical properties of each case was selected in such a way to ensure compressive crushing failure in the throat area with variable $\sqrt{A_c/A_h}$ ratios. This is intended to determine the increase in strength of the confined area due to this confinement from the larger column. A 70 × 70 mm square hinge, with 30 mm height, was used in all specimens. The area of the column was changed accordingly to provide different confinement levels. The factor $\sqrt{A_c/A_h}$ ranged between 2 and 4 throughout the case.

A 300 × 300 mm square footing is used in all specimens, with 300 mm height, and reinforced by one layer of 4ϕ14 in both directions. The concrete compressive strength of specimens CS-22, CS-32, and CS-40 were 22 MPa, 32 MPa and 40 MPa respectively, and the maximum aggregative size for concrete mix was 10 mm. The average measured yield strength of the longitudinal rebar was 520 MPa. The typical dimensions and reinforcing details of the specimens are shown in Table 1 and Fig. 1.

2.3 Test Setup

The load-displacement curve was determined using an automated testing machine provided with data acquisition unit. On line measurements of load and deflection were

Table 1. Characteristics of specimens

Specimen	f'_c (MPa)	Square column				Square hinge				$\sqrt{A_c/A_h}$
		Width (mm)	Height (mm)	Reinf. (mm^2)	Steel ratio	Width (mm)	Thick. (mm)	Reinf. (mm^2)	Steel ratio	
CS14-22	22	140	140	4ϕ12	2.29%	70	30	4ϕ10	6.41%	2.00
CS20-22	22	200	200	4ϕ14	1.53%	70	30	4ϕ10	6.41%	2.86
CS21-22	22	210	210	8ϕ12	2.03%	70	30	4ϕ10	6.41%	3.00
CS23-22	22	230	230	8ϕ12	1.69%	70	30	4ϕ10	6.41%	3.29
CS25-22	22	250	250	8ϕ12	1.43%	70	30	4ϕ10	6.41%	3.57
CS28-22	22	280	280	8ϕ14	1.56%	70	30	4ϕ10	6.41%	4.00
CS14-32	32	140	140	4ϕ12	2.29%	70	30	4ϕ10	6.41%	2.00
CS20-32	32	200	200	4ϕ14	1.53%	70	30	4ϕ10	6.41%	2.86
CS21-32	32	210	210	8ϕ12	2.03%	70	30	4ϕ10	6.41%	3.00
CS23-32	32	230	230	8ϕ12	1.69%	70	30	4ϕ10	6.41%	3.29
CS25-32	32	250	250	8ϕ12	1.43%	70	30	4ϕ10	6.41%	3.57
CS28-32	32	280	280	8ϕ14	1.56%	70	30	4ϕ10	6.41%	4.00
CS14-40	40	140	140	4ϕ12	2.29%	70	30	4ϕ10	6.41%	2.00
CS20-40	40	200	200	4ϕ14	1.53%	70	30	4ϕ10	6.41%	2.86
CS21-40	40	210	210	8ϕ12	2.03%	70	30	4ϕ10	6.41%	3.00
CS23-40	40	230	230	8ϕ12	1.69%	70	30	4ϕ10	6.41%	3.29
CS25-40	40	250	250	8ϕ12	1.43%	70	30	4ϕ10	6.41%	3.57
CS28-40	40	280	280	8ϕ14	1.56%	70	30	4ϕ10	6.41%	4.00

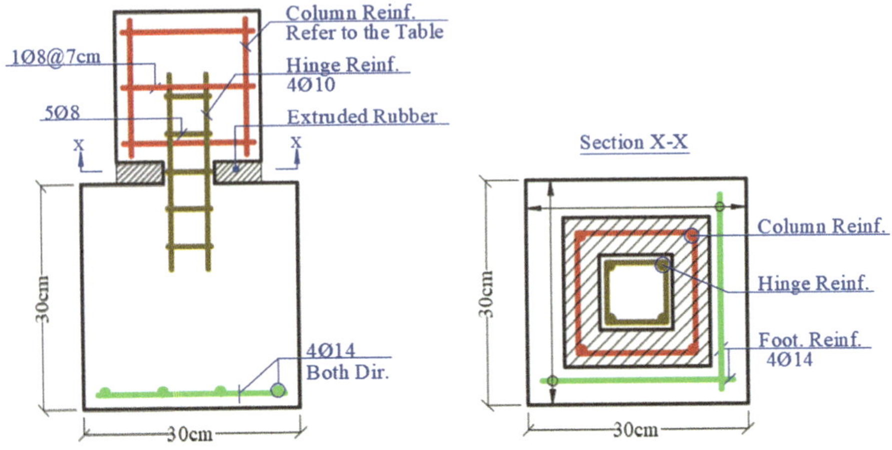

Fig. 1. Typical specimen detail

taken from transducers and transferred to the P.C. through the data acquisition unit. Specialized software allowed the on-line monitoring of the load deflection curve. Very large number of readings per second can be performed allowing an accurate and smooth transition of the curve. The load was acting monotonically at the top of the specimen through a displacement control method. The load produces axial deformation of the specimen until it fails. Figure 2 illustrates the description of the test.

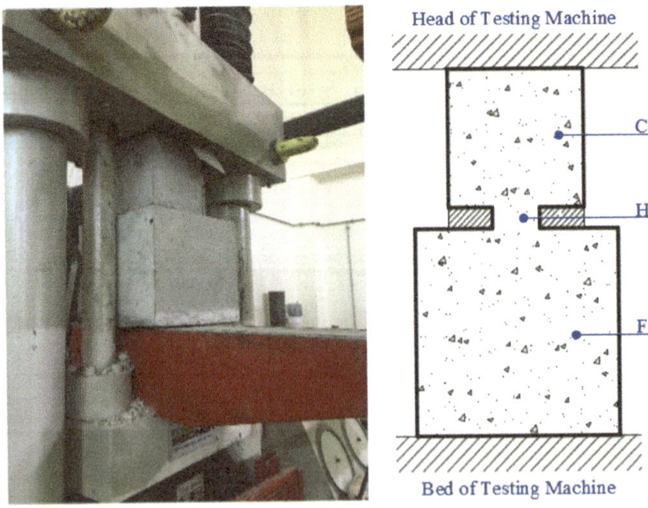

Fig. 2. Details of the specimen instrumentations, supports, and testing machine

2.4 Test Result

All specimens were subjected to a uniform axial load at the head of the column. The load deflection curve is then plotted online allowing monitoring of the behavior in the elastic and damaged zones. A comparison was made among the behaviors of the specimens.

2.4.1 Specimens CS-22

Figure 3 shows the 1st series which consists of six specimens having the same compressive strength, (f'_c = 22 MPa), and the square root of column area to hinge area ranged between 2 and 4.

Fig. 3. Specimens CS-22

Figure 4 presents the load displacement graphs for the CS-22. The specimens reached the ultimate capacity for a displacement ranged between 3.3 mm and 6.1 mm. Except for CS14-22, the specimens exhibited ductile failure modes, although the hinge failed way before 20 mm displacement. The failure type of the specimens is the concrete hinge throat crushing failure as shown in Fig. 5. This was obvious due to the reduction of the area of the hinge. Thus, the column is ultimately stronger than the hinge section. However, the confinement effect increased the strength of the hinged part. Most of the cases reached approximatively the same capacity, except for CS14-22. This is because the confinement factor, $\sqrt{A_c/A_h}$, of CS14-2 is 2, while it lies between 3 and 4 for the rest. This behavior is clarified next in Sect. 3.2.

Specimen CS14-22: The axial load is carried first by the hinge section, and the specimen exhibited ductile failure behavior. When the load became 247 KN, the plastic zone started, the cracks appeared at the hinge region, and hinge gap started to close. After reaching the maximum capacity at 346 KN and a displacement of 3.3 mm, the hinge region is extensively deteriorated. A wide crack appeared at the column edge and concrete started to spall at the corner. The test is then stopped.

Specimen CS20-22: Visible cracks were seen at the hinge region when the specimen reached a load of 331 KN. Then, the specimen entered into the plastic zone, and reached the maximum capacity at 441 KN with a displacement of 5.9 mm. The gap started to close. Meanwhile, longitudinal cracks formed at the base of the column and developed upward inducing concrete crushing. Then, the load was transferred to the

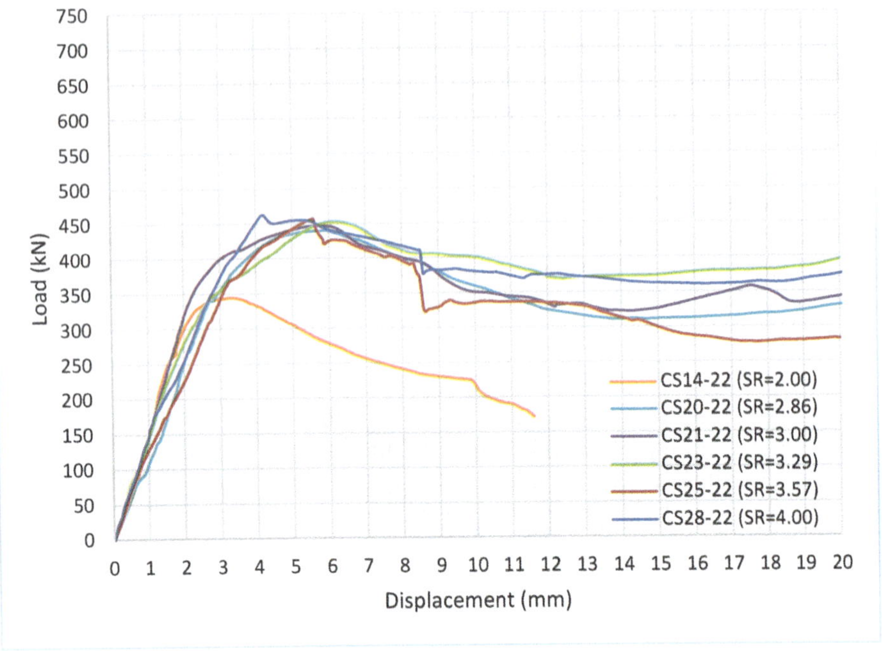

Fig. 4. Specimens CS-22 load displacement curves

footing and the test was stopped at a displacement of 20 mm exhibiting a ductile failure mode.

Specimen CS21-22: This specimen exhibited nearly the same behavior of CS20-22. The plastic zone started when the load was 309 KN. The specimen reached the maximum capacity of 448 KN with a displacement of 5.6 mm. the process continued, the hinge was fully closed. The load is then transferred to the column, and the test was stopped.

Specimen CS23-22: The specimen exhibited ductile failure behavior. When the load became 346 KN, the plastic zone started, and the cracks formed at the hinge region. The specimen reached the maximum capacity at 454 KN with a displacement of 6.1 mm. First hinge spalling was seen when the cracks were developed at the base of the column then longitudinally propagated upward. After that, the test was stopped while the deterioration was quite substantial.

Specimen CS25-22: This specimen exhibited a ductile failure behavior. The plastic zone started under a load of 368 KN. As the plastic zone continues, the specimen reached the maximum capacity of 458 KN with a displacement of 5.6 mm. the test was stopped when the cracks became extensively developed allowing the concrete to spall at the edge.

Specimen CS28-22: As CS25-22, the specimen exhibited a ductile failure behavior. Visible cracks were seen at the hinge region when the axial load is 395 KN. After the specimen reached the maximum capacity of 464 KN with a displacement of 4.2 mm,

Fig. 5. Crushing behavior of specimens CS-22

the hinge gap is closed and the concrete start to spall from the base of the column as well as the top of the footing. This is because both the column and the footing have nearly the same section.

2.4.2 Specimens CS-32

Figure 6 shows the 2nd series which consists of six specimens having the same compressive strength, (f'_c = 32 MPa), and the square root of column area to hinge area ranged between 2 and 4.

Figure 7 presents the load displacement graphs for CS-32. The specimens reached the ultimate capacity for a displacement ranged between 2.8 mm and 4.8 mm. Except for CS14-32, the specimens exhibited ductile failure modes, although the hinge failed way before 20 mm displacement. The failure type of the specimens is the concrete hinge throat crushing failure as shown in Fig. 8. This was obvious due to the reduction of the area of the hinge. Thus, the column is ultimately stronger than the hinge section. However, the confinement effect increased the strength of the hinged part. Same as series CS-22, most of specimens CS-32 reached approximatively the same capacity, except for CS14-32.

Fig. 6. Specimens CS-32

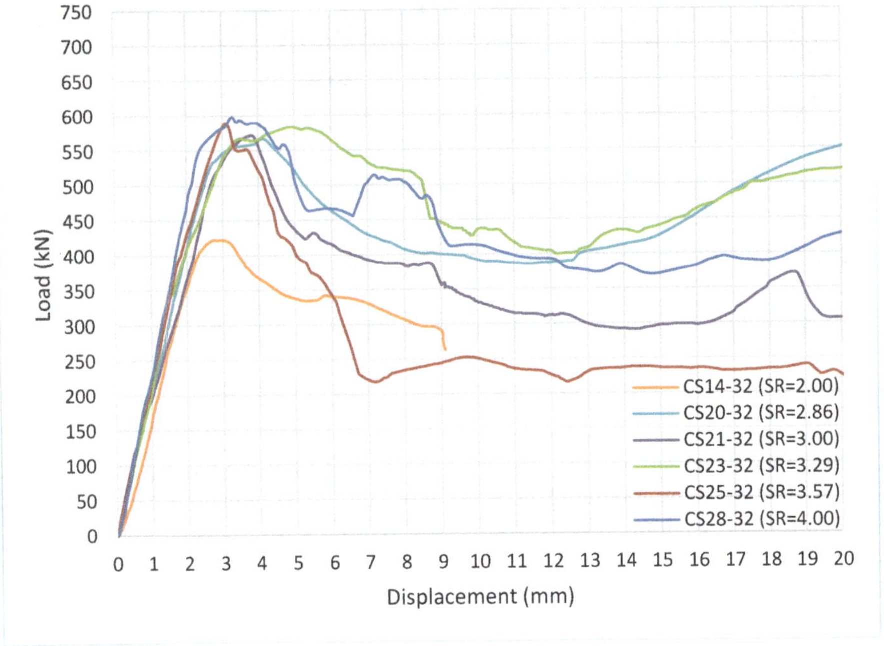

Fig. 7. Specimens CS-32 load displacement curves

Specimen CS14-32: The specimen exhibited the same behavior of CS14-22. The cracks appeared first at the hinge region once the load was 393 KN, thereafter the cracks propagated vertically from the bottom face of the column towards the top. The specimen reached the maximum capacity at 423 KN with a displacement of 2.9 mm. Then, the hinge deterioration is substantially increased, and subsequently failed at a displacement of 9 mm.

Specimen CS20-32: The specimen exhibited ductile failure behavior. As the plastic zone started, the cracks developed at the hinge region when the load is 511 KN. After reaching the maximum capacity at 567 KN at a displacement of 4.2 mm, the longitudinal

Fig. 8. Crushing behavior of specimens CS-32

cracks are wide enough, and the hinge failed at a displacement of 8.6 mm, and became no longer maintained to carry out the load. The hinge gap is nearly closed, and the cracks appeared at the edge of the footing. The test was stopped right after the concrete cover was crashed out of the footing.

Specimen CS21-32: Visible cracks were seen at the hinge region at a load of 516 KN. As the load increases, the plastic zone continued, and the cracks appeared at the bottom face of the column. The specimen reached the ultimate capacity of 572 KN at a displacement of 3.8 mm. after that, the load dropped suddenly, and a wide cracks was formed at the edge of the column and at the corner of the footing. The test was stopped right after the concrete cover at the corner was spalled out of the footing.

Specimen CS23-32: The specimen exhibited ductile failure behavior. The first crack appeared at the hinge region once the axial load is 536 KN. As the load increases, the hinge gap started to close. The specimen reached the ultimate capacity of 584 KN at a displacement of 4.8 mm. Thereafter, a large crack developed at the bottom of the column and extended extensively towards the top. Hereupon, the test was stopped.

Specimen CS25-32: The specimen exhibited ductile failure behavior. When the load reached 547 KN, the plastic zone started. The first crack at the hinge region appeared at 561 KN. As the plastic zone continued, the stiffness reduction caused by the cracking of concrete became apparent. The specimen reached the ultimate capacity of 590 KN at a displacement of 3.1 mm. At this time, the load is suddenly dropped, and a large crack was formed at the top of the footing showing that the deterioration is quite substantial.

Specimen CS28-32: Visible cracks were seen at the hinge region at an axial load of 558 KN. As the load increases, the specimen reached the ultimate capacity of 599 KN at a displacement of 3.3 mm. Hereafter, the hinge was nearly closed, and the cracks are extended upward to the column and downward to the footing. By the time, the cracks became obviously large inducing stiffness degradation, and the test was stopped while the specimen experienced large permanent displacement.

2.4.3 Specimens CS-40

Figure 9 shows the 3rd series which consists of six specimens having the same compressive strength, (f'_c = 40 MPa), and the square root of column area to hinge area ranged between 2 and 4.

Fig. 9. Specimens CS-40

Figure 10 presents the load displacement graphs for the CS-40. The specimens reached the ultimate capacity for a displacement ranged between 3.4 mm and 5.0 mm. Except for CS14-40, the specimens exhibited ductile failure modes, although the hinge failed way before 20 mm displacement. The failure type of the specimens is the concrete hinge throat crushing failure as shown in Fig. 11. This was obvious due to the reduction of the area of the hinge. Thus, the column is ultimately stronger than the hinge section. However, the confinement effect increased the strength of the hinged part. Same as other series, most of specimens CS-40 reached approximatively the same capacity, except for CS14-32.

Specimen CS14-40: The plastic phase started at a load 456 KN. The cracks appeared first at the hinge region once the load was 463 KN. Thereafter, the cracks propagated vertically from the bottom face of the column towards the top. The specimen reached the maximum capacity at 491 KN with a displacement of 3.7 mm. Then, the load which stayed at a constant rate of decreasing before dropping. At the end, the specimen reached a maximum deflection value of 7 mm.

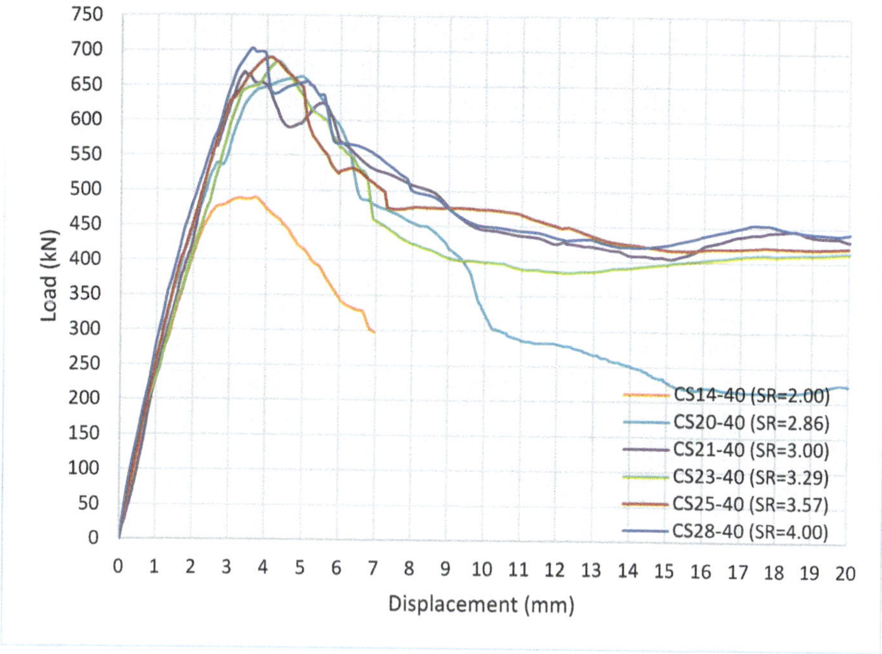

Fig. 10. Specimens CS-40 load displacement curves

Specimen CS20-40: The specimen exhibited ductile failure behavior. Under a load of 558 KN, visible cracks were formed at the hinge region. At this point the elastic phase finished, and the plastic zone started. As the load increases, the specimen reached the maximum capacity at 664 KN with a displacement of 5 mm. At this stage the hinge became closed and the specimen dropped but experienced large permanent displacement.

Specimen CS21-40: The specimen exhibited ductile failure behavior. The plastic phase started under a load of 563 KN, and the first crack appeared at the hinge region under an axial load was 572 KN. As the plastic zone continued, the cracks developed at the base of column and extended upward. The specimen reached the ultimate capacity of 671 KN at a displacement of 3.4 mm. Meanwhile, the cracks became obviously large inducing stiffness degradation. At the end, the test was stopped at 20 mm.

Specimen CS23-40: A hinge crack appeared first under a load of 628 KN. Under a load of 641 KN, the plastic phase started. As the load increases, the specimen reached the maximum capacity at 685 KN with a displacement of 4.4 mm. Meanwhile, the cracks developed at the base of the column and extended longitudinally upward. After that, the test was stopped right after the concrete cover at the corner was spalled out of the column.

Specimen CS25-40: The specimen exhibited ductile failure behavior. As of the axial load reached 629 KN, the plastic phase started, and the first hinge crack was seen under a load of 647 KN. As the plastic zone continues, the gap started to close, the specimen

Fig. 11. Crushing behavior of specimens CS-40

reached the maximum capacity at 691 KN with a displacement of 4.1 mm. Meanwhile, the cracks were developed at the base of the column then longitudinally propagated upward.

Specimen CS28-40: The specimen exhibited nearly the same behavior of CS25-40. Visible cracks were seen at the hinge region at an axial load of 669 KN. The specimen reached the ultimate capacity of 704 KN at a displacement of 3.6 mm. Meanwhile, the cracks were developed at the base of the column then longitudinally propagated upward. The test was stopped while the specimen experienced large permanent displacement.

3 Strength Calculation

3.1 Column Axial Strength

Based on ACI-318-14 code, the ultimate capacity (P_{ul}) and bearing strength ($øP_n$) of the columns were determined using Eqs. 1a and 1b, where A_g is the area of the column, A_s is the area of steel reinforcement, f'_c is the 28-days concrete compressive strength, and

f_y is the steel yield tensile strength (f_y = 520 MPa) [11]. Table 2 presents the axial strengths of all specimens.

$$P_{ul} = 0.85f_c'(A_g - A_{st}) + f_y A_{st} \tag{1a}$$

$$\phi P_n = 0.8\phi P_{ul} \tag{1b}$$

3.2 Hinge Axial Strength

The ultimate capacity (P_{ulh}) and bearing strength ($øP_{nh}$) of the concrete hinge were determined using Eqs. 2a and 2b. ACI 318-14 code limits the amplification of the bearing capacity provided by confinement from larger column on the smaller hinge area $\sqrt{A_c/A_h}$ to 2. The code indicates that even if this ratio exceeds 2, it should be taken as 2 [11]. This is because the confining effect from the larger area reaches an asymptotic value where any increase in the area of the column would cause negligible increase in the ultimate capacity of the hinge assembly.

$$P_{ulh} = 0.85f_c'(A_g - A_{st})\sqrt{A_c/A_h} + f_y A_{st} \tag{2a}$$

$$\phi P_{nh} = 0.8\phi P_{ulh} \tag{2b}$$

Table 2. Column axial strength

Specimen	f_c' (MPa)	Square column			$øP_n$ (KN)	P_{ul} (KN)
		Width (mm)	Reinf. (mm²)			
CS14-22	22	140	4ϕ12	448	307	591
CS20-22	22	200	4ϕ14	612	548	1,055
CS21-22	22	210	8ϕ12	896	662	1,274
CS23-22	22	230	8ϕ12	896	748	1,438
CS25-22	22	250	8ϕ12	896	841	1,618
CS28-22	22	280	8ϕ14	1,224	1,081	2,080
CS14-32	32	140	4ϕ12	448	392	754
CS20-32	32	200	4ϕ14	612	723	1,390
CS21-32	32	210	8ϕ12	896	853	1,641
CS23-32	32	230	8ϕ12	896	978	1,880
CS25-32	32	250	8ϕ12	896	1,114	2,142
CS28-32	32	280	8ϕ14	1,224	1,423	2,736
CS14-40	40	140	4ϕ12	448	460	884
CS20-40	40	200	4ϕ14	612	862	1,657
CS21-40	40	210	8ϕ12	896	1,006	1,935
CS23-40	40	230	8ϕ12	896	1,162	2,234
CS25-40	40	250	8ϕ12	896	1,331	2,560
CS28-40	40	280	8ϕ14	1,224	1,695	3,260

The code indicates explicitly that this amplification should be used in case of confinement from all directions. However, many practicing engineers and practical design methods consider Eq. 2a and 2b for one way hinges design. This may be justified as the one way confinement from the one way hinges shouldn't be disregarded. The hinge axial strength are shown in Table 3.

3.3 Confinement Effects on the Ultimate Capacity of the Hinge

Figure 12 summarizes the ultimate two-way hinge capacity with the variation of the ratio of area of column to hinge $\sqrt{A_c/A_h}$. The magenta thick square dot line presents the code recommended bearing strength of the throated area ($øP_{nh}$) due to the confinement ($\sqrt{A_c/A_h} = 2$). The blue dash dot presents the code ultimate capacity of the throat (P_{ulh}). However, the red long dash presents the modified ultimate capacity of the throat taking into account the confinement level ($\sqrt{A_c/A_h}$) up to 3. The green solid line presents the ultimate capacity of the hinges considered in the test.

Table 3. Hinge axial strength

Specimen	f'_c (MPa)	Square hinge Width (mm)	Thickness (mm)	Reinf. (mm^2)		$\sqrt{A_c/A_h}$	$øP_{nh}$ (KN)	P_{hul} (KN)
CS14-22	22	70	30	4ϕ10	314	2.00	174	335
CS20-22	22	70	30	4ϕ10	314	2.86	174	335
CS21-22	22	70	30	4ϕ10	314	3.00	174	335
CS23-22	22	70	30	4ϕ10	314	3.29	174	335
CS25-22	22	70	30	4ϕ10	314	3.57	174	335
CS28-22	22	70	30	4ϕ10	314	4.00	174	335
CS14-32	32	70	30	4ϕ10	314	2.00	215	413
CS20-32	32	70	30	4ϕ10	314	2.86	215	413
CS21-32	32	70	30	4ϕ10	314	3.00	215	413
CS23-32	32	70	30	4ϕ10	314	3.29	215	413
CS25-32	32	70	30	4ϕ10	314	3.57	215	413
CS28-32	32	70	30	4ϕ10	314	4.00	215	413
CS14-40	40	70	30	4ϕ10	314	2.00	247	475
CS20-40	40	70	30	4ϕ10	314	2.86	247	475
CS21-40	40	70	30	4ϕ10	314	3.00	247	475
CS23-40	40	70	30	4ϕ10	314	3.29	247	475
CS25-40	40	70	30	4ϕ10	314	3.57	247	475
CS28-40	40	70	30	4ϕ10	314	4.00	247	475

Based on the test results, the two-way hinge capacity didn't have the same trend as what was recommended by the American Concrete Institute (ACI). The ultimate capacity of the two way hinges tended to increase up the confinement factor, called CF,

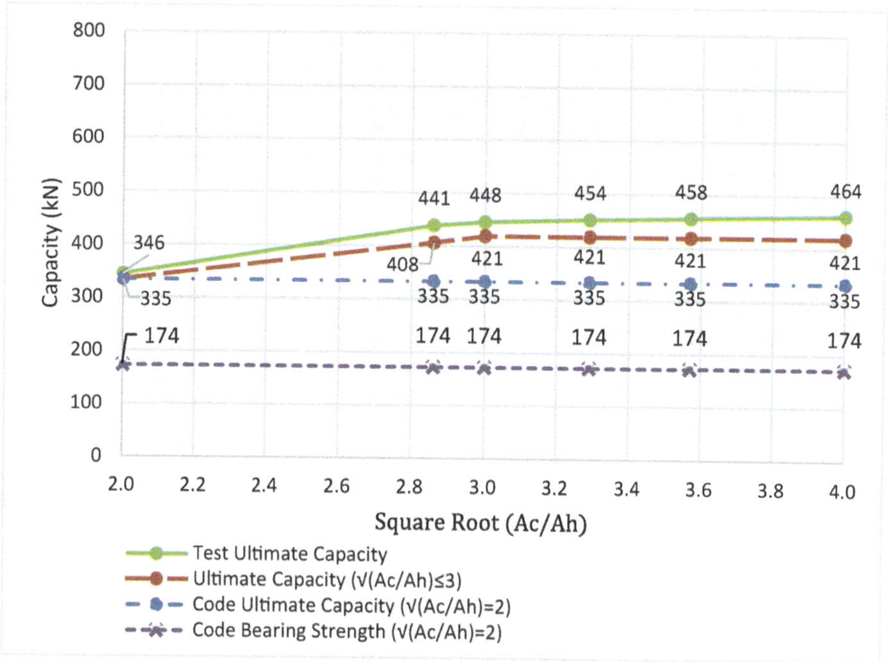

(a) Specimens CS-22

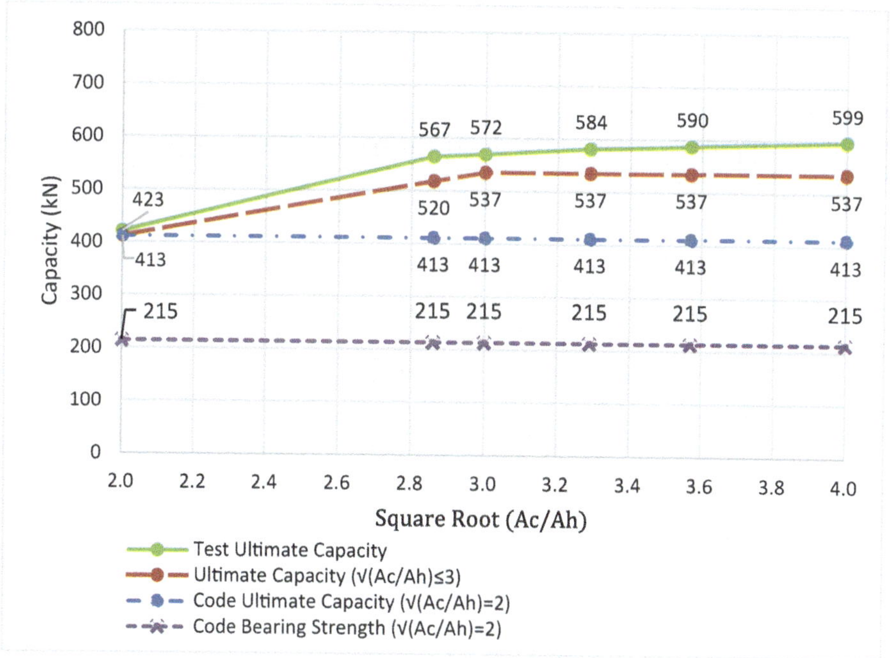

(b) Specimens CS-32

Fig. 12. Square root of area ratio versus hinge ultimate capacity (a) Specimens CS-22, (b) Specimens CS-32, (c) Specimens CS-40

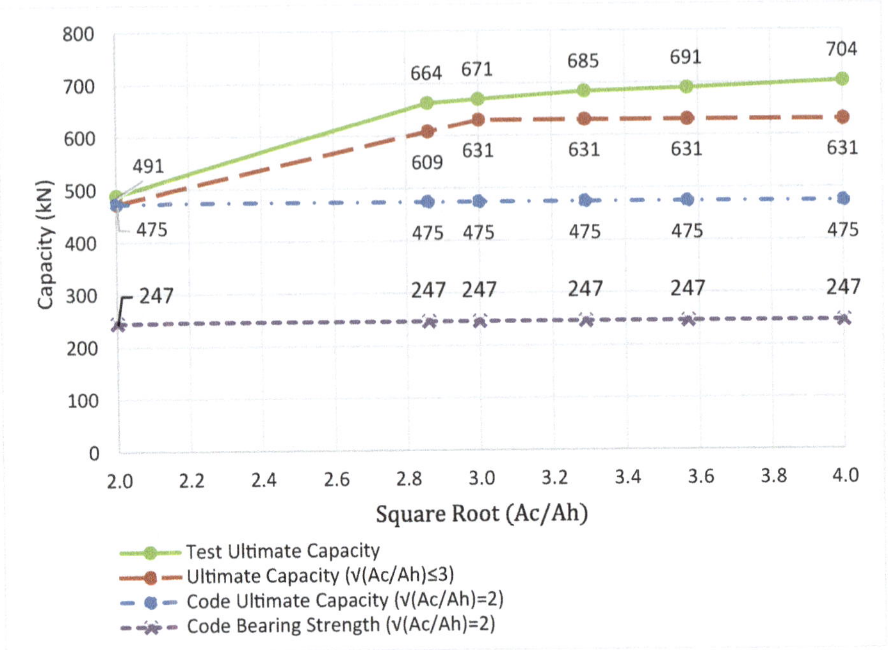

(c) Specimens CS-40

Fig. 12. *(continued)*

to 3. Obviously, the ultimate capacity of the hinge increased nearly 3.5 times at the maximum asymptotic value. Table 4 shows the confinement factor "CF" for the specimens considered.

Table 4. Hinge Confinement Factor (CF)

Specimen	$\sqrt{A_c/A_h}$	Code bearing strength $\sqrt{A_c/A_h} = 2$ (KN)	Code ultimate capacity $\sqrt{A_c/A_h} = 2$ (KN)	Ultimate capacity $\sqrt{A_c/A_h} \leq 3$ (KN)	Test results (KN)	CF
CS14-22	2.00	174	335	335	346	2.13
CS20-22	2.86	174	335	408	441	3.24
CS21-22	3.00	174	335	421	448	3.32
CS23-22	3.29	174	335	421	454	3.39
CS25-22	3.57	174	335	421	458	3.44
CS28-22	4.00	174	335	421	464	3.51
CS14-32	2.00	215	413	413	423	2.08
CS20-32	2.86	215	413	520	567	3.24

(continued)

Table 4. (*continued*)

Specimen	$\sqrt{A_c/A_h}$	Code bearing strength $\sqrt{A_c/A_h} = 2$ (KN)	Code ultimate capacity $\sqrt{A_c/A_h} = 2$ (KN)	Ultimate capacity $\sqrt{A_c/A_h} \leq 3$ (KN)	Test results (KN)	CF
CS21-32	3.00	215	413	537	572	3.28
CS23-32	3.29	215	413	537	584	3.37
CS25-32	3.57	215	413	537	590	3.42
CS28-32	4.00	215	413	537	599	3.49
CS14-32	2.00	247	475	475	491	2.10
CS20-32	2.86	247	475	609	664	3.21
CS21-32	3.00	247	475	631	671	3.26
CS23-32	3.29	247	475	631	685	3.35
CS25-32	3.57	247	475	631	691	3.38
CS28-32	4.00	247	475	631	704	3.47

4 Conclusion

This paper presents the experimental analysis conducted on several two-way hinges to assess their behavior on concrete structures with specific focus on the confinement effect, of the square root of the column area over the hinge area, on the ultimate capacity of the hinge. Eighteen two-way hinges were tested in laboratory facilities located at Beirut Arab University. Based on the tests reported, several conclusions can be made.

(a) Two-way hinges with high square root of column area to hinge area ratio ($\sqrt{A_c/A_h}$) exhibited ductile failure modes.

(b) Two-way hinge capacity didn't have the same trend as what was recommended by the American Concrete Institute (ACI). The ultimate capacity of the two way hinges tended to increase up the confinement factor ($\sqrt{A_c/A_h}$) to 3. Obviously, the ultimate capacity of the hinge increased nearly 3.5 times at the maximum asymptotic value.

(c) Despite the confinement effect increased the strength of the hinged part, however, the main failure type of the specimens is the concrete hinge throat crushing failure. This was obvious due to the reduction of the area of the hinge.

Acknowledgements. The authors wish to express their gratitude to BAU Department of Civil and Environmental Engineering for their support.

References

1. Cheng, Z., Saiidi, M.S., Sanders, D.H.: Development of a seismic design method for reinforced concrete two-way bridge column hinges. Executive summary report, June 2006
2. Jiang, Y., Saiidi, M.: Response and design of R/C one-way pier hinges in strong direction. ASCE J. Struct. Eng. **121**(8), 1236–1244 (1995). ASCE Reston, VA
3. McLean, D.I., Lim, K.Y.: Moment-reducing hinge details for the bases of bridge columns. Washington State Department of Transportation, Final report, October 1990
4. Mortensen, J., Saiidi, M.: A performance-based design method for confinement in circular columns. Report No. CCEER-02-7, Center for Civil Engineering Earthquake Research, Department of Civil Engineering, University of Nevada, Reno, Nevada, November 2002
5. Zaghi, A.E., Saiidi, M.S.: Seismic design of pipe-pin connections in concrete bridges columns. J. Earthq. Eng. **14**, 1253–1302 (2010)
6. Linke, C., Pekcan, G., Itani, A.: Detailing of seismically resilient special truss moment frames. Report No. CCEER-09-02, Center for Civil Engineering Earthquake Research, Department of Civil and Environmental Engineering, University of Nevada, Reno, Nevada, August 2009
7. Hills, D., Saiidi, M.: Design, construction, and nonlinear dynamic analysis of three bridge bents used in a bridge system test. Report No. CCEER-09-03, Center for Civil Engineering Earthquake Research, Department of Civil and Environmental Engineering, University of Nevada, Reno, Nevada, August 2009
8. Mehrsoroush, A., Saiidi, M.S.: Earthquake-resistant telescopic pipe pin column base connections for accelerated bridge construction. In: Tenth U.S. National Conference on Earthquake Engineering, Frontiers of Earthquake Engineering, Anchorage, Alaska, July 2014
9. Zaghi, A., Saiidi, M.: Seismic performance of pipe-pin two-way hinges in concrete bridge columns. ASCE J. Bridge Eng. **16**(3), 340–350 (2011)
10. Marx, S., Schacht, G.: Concrete hinges in bridge engineering. Eng. Hist. Heritage **168**(EH2), 65–75 (2015)
11. Building code requirements for structural concrete (ACI 318M-14) and commentary (ACI 318RM-14), American Concrete Institute

Post Buckling Behavior of Steel Plate Girder Panels Under Shear Loading

Hassan Ghanem, Safwan Chahal(✉), Wafik Ajam, and Adel Kurdi

Beirut Arab University, Beirut, Lebanon
h.ghanem@bau.edu.lb, safwanchahal79@gmail.com

Abstract. Previously, the shear design of web and transverse stiffeners was based on the initial shear buckling in the web without the tension field action. The American Institute of Steel Construction (AISC) adopted post-buckling tension field strength into its specifications for stiffened interior web panels but exclude the post-buckling tension field action for the end web panel. In this study, a finite element method (FEM) analysis on an isolated panel confirms the view that post-buckling strength of steel plate girders is attributed to a non-uniform shear stress distribution along the boundary of the plates, varying from the critical stress in one corner up to the shear-yield stress in the tension corner with no need for any diagonal tension. It was also shown that the presence of flanges with bending strength does rise the shear capacity in the panel. However; light flanges give a slight increase in shear resistance without diagonal tension, but heavy flanges are shown to be capable of developing true diagonal tension, leading to a significant gain in total shear capacity.

Keywords: Shear stress · Buckling · Tension field action

1 Introduction

When a web is subjected to shear, before it buckles, equal tensile and compressive principal stresses are developed within the plate. If the applied loading is increased, the buckling of the plate does not limit the shear resistance [1–4].

The post buckling strength was first attributed by Wagner in 1929 to a complete uniform tension field reacted by rigid flanges [5]. Basler in 1961 observed that in welded plate girders, the flanges are too flexible to resist the web tension, and he proposed a partial tension field anchored against the vertical edges of the panel [6]. The Wagner and Basler models represent the two extreme cases of the many models that have been proposed that make use of the idea of diagonal tension. Marsh suggested a different model in which, after initial buckling, the uniform shear distribution along the boundary changes, with increasing shear stress toward the tension corner, but with no stress normal to any boundary [7].

Marsh and Ajam use the finite element approach and obtained a numerical solution that can be formulated in terms of simple physical concepts [8]. A classical stability analysis for the critical shear stress in a web is based on a uniform shear-stress distribution. In practice, for a thin web, a uniform distribution of the stress is not possible if there are initial imperfections [9–11]. As the buckled plates deflect further, the non-

H. Rodrigues et al. (Eds.): GeoMEast 2019, SUCI, pp. 147–159, 2020.
https://doi.org/10.1007/978-3-030-34216-6_10

uniformity stresses at the edges increase and the resistance to shear force is provided by pure shear stress along the boundaries, which, in the limit, varies from approximately the initial buckling stress at the compressive corners, to the yield stress in shear in the tension corners, where the short compressive strips are sufficiently stable [12–14].

An isolated panel subjected to shear distortion is capable of carrying a shear force well in excess of that causing initial shear buckling, even when the boundary flanges have no bending strength.

This study covers two conditions of end panels: 1-Shear capacity for a square panel with flanges that have area (axial stiffness) but no flexural stiffness. 2-Behavior and shear capacity of a square panel with flexural stiffness flanges.

2 Square Panels with Flanges That Have No Bending or Torsional Stiffness

The initial analysis is for a single square panel with stiffeners along all four edges, the stiffeners having an area, and hence axial stiffness, but no flexural rigidity (β). When this panel is subjected to a shear force, it is evident that there can be no normal stress at the boundaries and only shear stress can exist there.

ABAQUS program was used in the analysis. Rather than applying a shear loading, imposed displacements of the stiffeners are used. In Fig. 1, the distance OA′ is reduced, and distance OB is increased, by an amount δ, causing a shear displacement along the boundary, leading to a shear stress distribution which will be determined by the analysis.

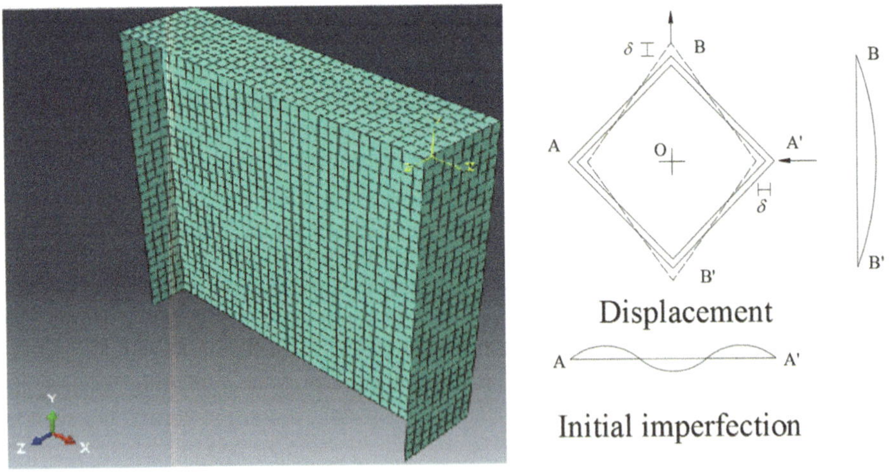

Fig. 1. A square plate finite element grid, 30 × 30 elements with light flanges.

For a square panel, of depth d and thickness t, subjected to a uniform shear distribution the elastic buckling stress, τ_{cr} is given by:

$$\tau_{cr} = 9.34\,\pi\,E\,t^2/(12(1-v)d^2) \tag{1}$$

Where v is the Poisson's ratio and E is the modulus of elasticity. The total shear force at initial elastic buckling is equal to:

$$V_{cr} = \tau_{cr} dt \tag{2}$$

After buckling, as the applied displacement increases, the shear stress is no longer uniform but increases in value towards the tension corner. This process is continued until the web yields in shear at the corner. The total shear force V, is given by the integration of the shear flux along the boundary. For more displacement of the flanges the extent of the yield zone increases, but for the case with flanges having no flexural stiffness, the actual total shear force along the boundary has been shown to decrease, due to a change in the stress distribution.

Figure 2 shows the variation of the central deflection w/t with respect to the shear displacements. Irrespective of the assumed initial central deflection, the central deflection increases only when the imposed shear displacement reaches a certain critical value τ_{cr}. This point is the bifurcation point and it agrees very well the linear buckling theory.

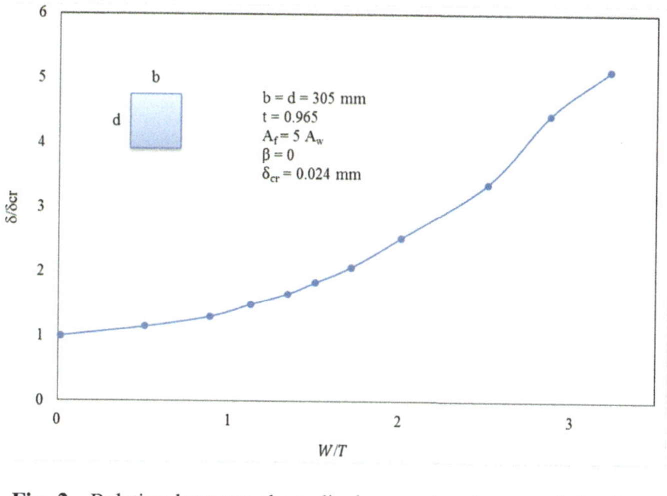

Fig. 2. Relation between shear displacement and center deflection

At the bifurcation point the plate deflection is formed by three half-waves in the compression direction and one half-wave in the tension direction. For non-dimensional presentation of the results, the value on the abscissa is the ratio $\Delta = \delta / \delta_y$, where δ_y is the nominal displacement when the shearing strain is τ_y/G given by:

$$\delta_y = (d/2\sqrt{2})(\tau_y/G) \tag{3}$$

Where G is the elastic shear modulus. The ordinate is V/V_y, where V is the total shear force and $V_y = \tau_y dt$.

Figure 3 shows the behavior of a panel with $d/t = 316$. t = 1 mm, and a flange area $A_f = 5A_w = 5dt$.

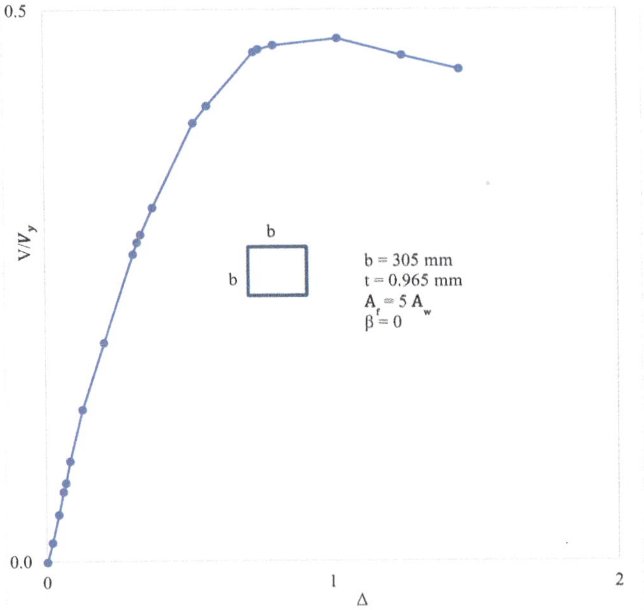

Fig. 3. Shear force V displacement

The variation of the shear stress along the boundary is shown in Fig. 4. The abscissa is the ratio x/b, and the ordinate is τ/τ_y.

Figure 5 shows the ratio of the central deflection to the thickness plotted along the compression diagonal, for various displacement values. As shown, the deflection is formed by three half waves in the compression direction.

The proposed model in square panel resembles two sets of diagonal strips, one in tension and in compression; limiting to τ_{cr} at the longest compression diagonal and τ_y at the longest tension diagonal. It is assumed that the buckling stress at the boundaries varies in parabolic form as shown in Fig. 6, the shear stress at a point on the edge is:

$$\tau_s = \tau_{cr} + (\tau_y - \tau_{cr})\,(x/d)^2 \tag{4}$$

In which x = distance from the corner; τ_{cr} = theoretical critical shear stress which occurs at x = Zero, and τ_y = yielding shear stress which occurs at x = d.

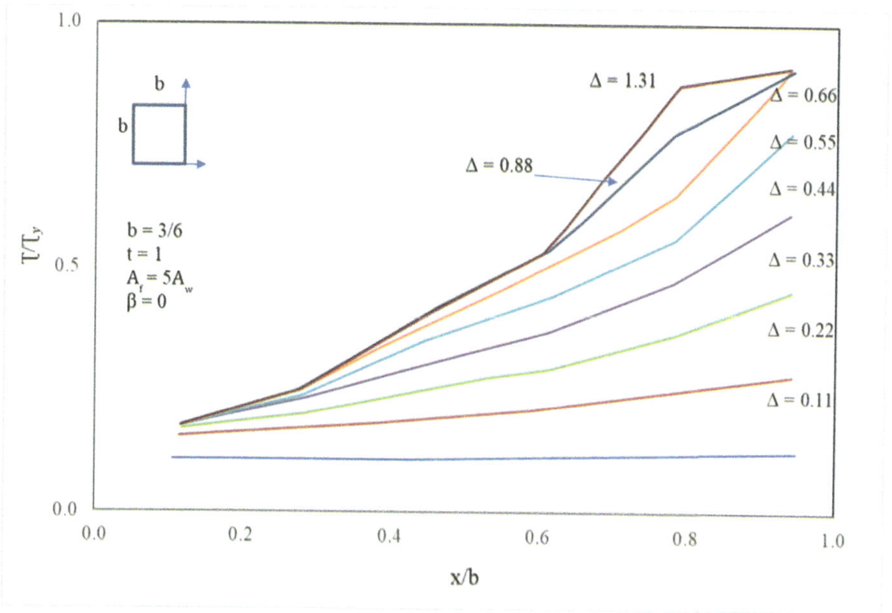

Fig. 4. Shear stress along boundary for varying displacement

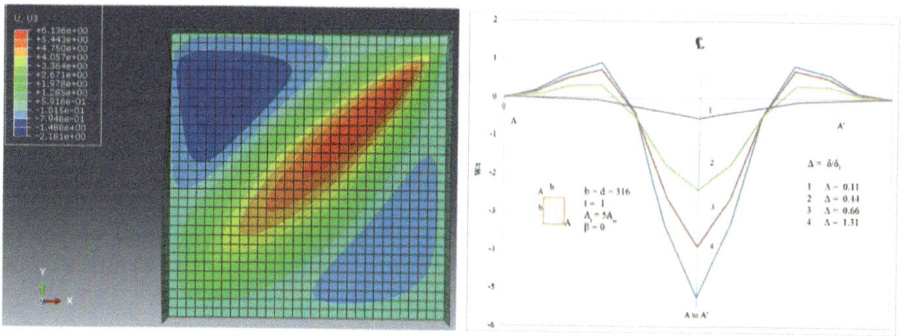

Fig. 5. Deflection along compression diagonal

The principal stress along any line running at 45° remains constant along that line, with a maximum value of $\sigma_c = \sigma_t = \tau_y$. At any point in the web, for values of $(x+y)<d, x<y$, the stresses are given by:

$$\sigma_c = \tau_{cr} + (\tau_y - \tau_{cr})\,((x-y)/d)^2 \tag{5}$$

$$\sigma_t = \tau_{cr} + (\tau_y - \tau_{cr})\,((x+y)/d)^2 \tag{6}$$

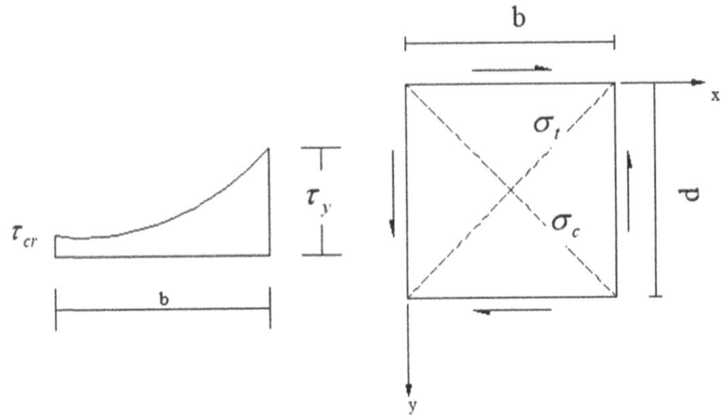

Fig. 6. Square panel under shear force in the proposed model

$$\tau_{xy} = \tau_{cr} + (1/2)\ (\tau_y - \tau_{cr})\ ((x-y)^2 + (x+y)^2)/d^2 \tag{7}$$

The shear force V on a boundary is the area under the shear stress - thickness curve.

$$V = t\int_0^d \tau_s dx \tag{8}$$

The critical shear stress depends upon the boundary conditions of the isolated panel, but the true boundary conditions for a girder web are difficult to establish accurately because the degree of restraint imposed by the flanges and by the adjacent web panels cannot be evaluated. It can be assumed conservatively that all the boundaries of the web panel are simply supported.

Up to first yield, Eq. 4 represents the variation of the shear stress along the boundaries. The shear forces in the plate at first yield in the web is obtained by integrating the shear stress distribution along the edge, giving:

$$V_w = (1/3)(2\tau_{cr} + \tau_y)dt \tag{9}$$

For more rigid flanges the ultimate shear force is equal to the shear force up to first yielding plus the contribution of the flanges in shear.

The failure load can be determined from a consideration of the mechanism developed in the frame panel. In this model, up to first yielding in shear there is no normal force at the boundary. As the load increases, σ_t will increase and σ_c will reduce in such way as to conserve the value of shear stress τ_y at the tension corner, as the normal force increases. The tension stress in the longer diagonal increases up to the value of σ_y, the final condition approaching a series of yielded diagonals strips that maximum value of the normal force developed at the boundary is t $\sigma_y/2$.

3 Square Panels with Flanges That Have Rigidity

A finite element analysis was conducted, for a panel with a high slenderness ratio with flexural rigidity as shown in Fig. 7. The two tension corners are free to move; one compression corner is fixed in the X and Y directions. The load is applied at the other compression corner in X and Y direction. The relative proportions of the flange are expressed by the ratio of flange area to web area A_f/A_w and the ratio β, of plastic moment strength of the flange $(0.25b_f\ t_f^2\ \tau_y)$ to that required to resist a tension field encompassing the whole panel, with an interior hinge at the midpoint $(c_y b^2 t/32)$, this $\beta = 8b_f\ t_f^2\ /b^2 t$, which is usually less than 0.1 for practical welded plate girders.

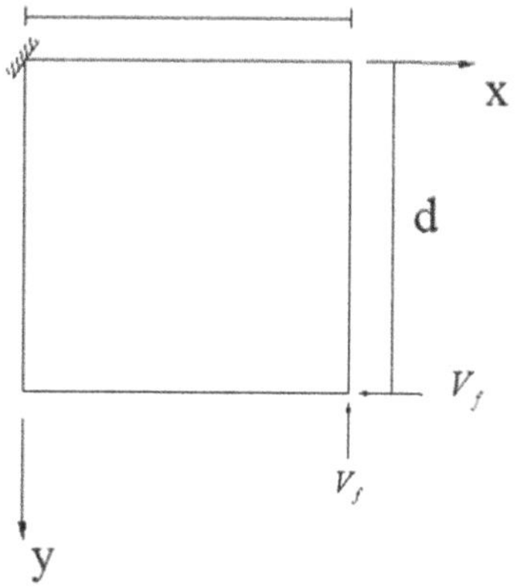

Fig. 7. Finite element test for square panel under compression load at one corner

After first yielding in shear at the tension corner, as the load increases the yielded zone spreads, and normal forces are created on the flanges. Failure occurs when the hinges have formed in the flanges which together with the yielded zone A A′ A″, B B′ B″ form a plastic mechanism as shown in Fig. 8.

By considering a rotation ø to occur at the plastic hinges producing the mechanism shown in Fig. 9, the work done in the plastic deformation of the flange and web is:

$$\begin{aligned} 2V_f c\phi &= 8M_p\phi + 4(\sigma_y/2)ctc\phi/2 \\ V_f &= 4M_p/c + ct\sigma_y/2 \end{aligned} \tag{10}$$

The load is applied at the compression corner in Fig. 9 in the Y direction only, and is assumed to be reacted by a uniformly distributed load σ_y t/2 normal to the flange exerted by the web. Since the plastic hinge will occur at the point of maximum bending

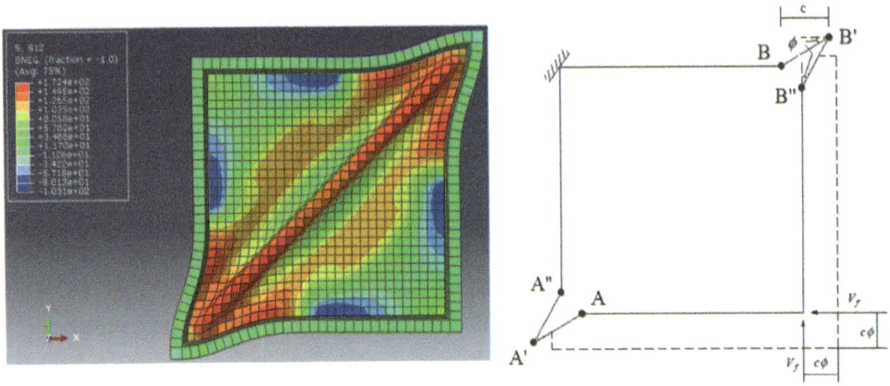

Fig. 8. Hinge formation in the flange of isolated panel

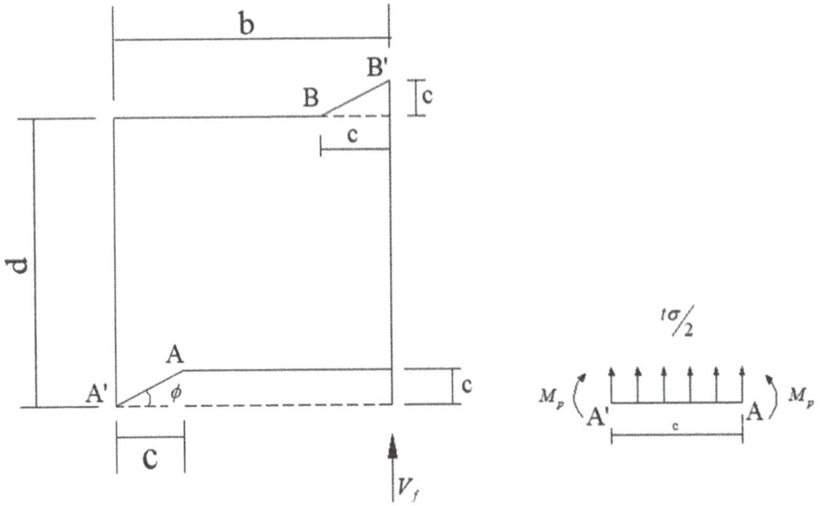

Fig. 9. Formation of two hinges in the flange

moment where the shear is zero, the position of the internal hinge A is obtained by considering the equilibrium of the beam section A A′. Taking moments about A′, one obtains:

$$
\begin{aligned}
&(ct\sigma_y/2)c/2 = 2\,M_p \Rightarrow 4M_p/c = ct\sigma_y/2 \\
&c = \sqrt{(8\,M_p(\sigma_y t)}
\end{aligned}
\tag{11}
$$

The failure load will be equal to the force at first yield plus the contribution of the flange when the normal forces are developed.

$$V_u = V_w + V_f \tag{12}$$

In the case of a girder, the internal hinges A″ and B″ will not be formed because we have a continuous web, and the failure load will be determined from a consideration of the mechanism developed in Fig. 9. Consider a rotation ø at the plastic hinges, the work is done only on the sections AA′ and B B′. For equal flanges, the distance c for upper and lower flanges will be equal, and the work done by the axial force in the flanges will be zero.

The virtual work done in the mechanism is:

$$\begin{aligned} V_f c\phi &= 4M_p\phi + [2(\sigma_y/2)(ct)/2](c\phi) \\ V_f &= 4M_p/c + (\sigma_y/2)(ct) \end{aligned} \tag{13}$$

Equation 13 is the same as Eq. 10. The first term of this equation, on the right-hand side, represents the frame resistance in shear, the second term represents the contribution of the flanges to increasing the web capacity. By using Eq. 11 these two terms are shown to be equal. Putting Eqs. 9 and 10 into Eq. 12 gives.

$$V_u = (1/3)(2\tau_{cr} + \tau_y)dt + \sigma_y ct \tag{14}$$

Using the value of c from Eq. 11 in Eq. 14 the ultimate shear force becomes:

$$V_u = (1/3)(2\tau_{cr} + \tau_y)dt + \sqrt{8M_p\sigma_y t} \tag{15}$$

It should be noted that the value of the plastic moment in the flange will be reduced by the presence of the axial force P′. The value of the effective moment M′ is given by:

$$M' = M_p(1 - (P'/P)^2) \tag{16}$$

Where P is the axial force to yield the flange, equal to $A_f\sigma_y$.

4 Extreme Cases

From Eq. 14, it can be seen that the shear capacity is composed of two components. The first component represents the capacity of the web up to first yield without any contribution from the flange; the second component represents the additional shear capacity of the flange bending rigidity.

In the case of a girder with very weak flanges, taking account of the reduced moment resistance due to the axial force in the flange, the value of the flange strength M_p becomes small and the value of c becomes very small; the second term of Eq. 13 becomes negligible so that.

$$V_u = (1/3)(2\tau_{cr} + \tau_y)dt \tag{17}$$

When the flanges are very strong, the distance of the plastic hinge from the end of the panel c increases, as shown by Eq. 11. When c because equal to the width b, the hinges form at the four corners of the panel to form a "picture frame" mechanism, as shown in Fig. 10.

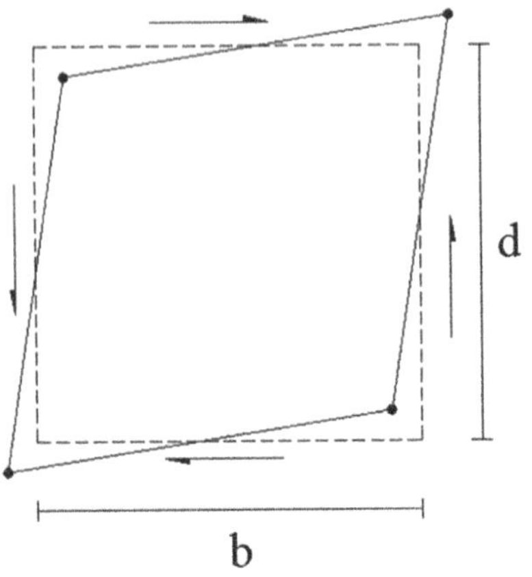

Fig. 10. Hinge formation at all corners of the panel

By substituting the value of c = b = d into Eq. 11 we get:

$$M_p = d^2\sigma_y t/8 \tag{18}$$

When determining the ultimate shear load for any girder, the flange strength should be evaluated to see whether it exceeds the limiting value given by Eq. 18. If this is so (it will not often be the case for girders of civil engineering proportions), the expression for shear capacity is:

$$V_u = ((1/3)(2\tau_{cr} + \tau_y) + \sigma_y)dt \tag{19}$$

5 Comparison of Predicted Shear Strength

Table 1 gives the properties of the plate girder panels. Table 2 gives comparisons of predicted shear strengths according to the present model with results of tests from a number of sources. Table 3 summarized the comparison of test results.

Table 1. Properties of square panels

Ref	Girder	t mm	d mm	A_w	t_f mm	b_f mm	A_f/A_w	β
15	TG1	2.72	609.8	1658.0	4.70	101.6	0.29	0.018
15	TG2	2.72	609.8	1658.0	6.55	101.6	0.40	0.034
15	TG14	.965	304.8	294.0	3.12	76.2	0.81	0.093
15	TG15	.965	304.8	294.0	5.00	76.2	1.30	0.224
15	TG16	.965	304.8	294.0	6.45	76.2	1.67	0.451
15	TG17	.965	304.8	294.0	9.32	76.2	2.41	0.850
15	TG18	.965	304.8	294.0	12.95	76.2	3.36	1.593
15	TG19	.965	304.8	294.0	15.52	76.2	4.02	2.004
15	TG20	2.03	304.8	619.0.	3.25	76.2	0.40	0.046
15	TG21	2.03	304.8	619.0.	4.88	76.2	0.60	0.097
15	TG22	2.03	304.8	619.0.	6.48	76.2	0.80	0.207
15	TG23	2.03	304.8	619.0.	9.22	76.2	1.14	0.419
15	TG24	2.03	304.8	619.0.	12.95	76.2	1.59	0.724
15	TG25	2.03	304.8	619.0.	15.54	76.2	1.91	0.914
15	G7T1	4.98	1270.0	6325.0	19.50	310.0	0.96	0.120
16	TG1′	2.50	1000.0	2500.0	5.06	160.0	0.32	0.019
16	TG2′	2.50	1000.0	2500.0	10.00	200.0	0.80	0.090
16	TG3′	2.50	1000.0	2500.0	16.43	200.0	1.31	0.241
16	TG4′	2.50	1000.0	2500.0	20.16	200.0	1.61	0.366
16	TG5′	2.50	1000.0	2500.0	29.73	200.0	3.00	0.790
17	PC3	1.0	800.0	800.0	10.0	250.0	3.12	0.431

Table 2. Comparison between test results, P.D.P, Hoglund [18] and Rockey [15]

Girder	d/t	σ_y MPa	σ_{yf} MPa	V P.D. P. KN	V Hoglund KN	V Rockey KN	V Exp KN
TG1	224	253	253	120.0	140.0	143.0	120.0
TG2	224	253	253	130.0	138.0	142.0	126.0
TG14	361	219	309	24.0	21.0	23.0	25.0
TG15	.361	219	289	29.0	24.0	26.0	29.0
TG16	361	219	249	35.0	29.0	30.0	31.0
TG17	.361	219	315	43.0	37.0	36.0	39.0
TG18	361	219	306	54.0	46.0	44.0	51.0
TG19	.361	219	268	59.0	50.0	48.0	55.0
TG20	150	229	.309	65.0	64.0	68.0	51.0
TG21	150	229	289	73.0	67.0	71.0	71.0
TG22	150	229	.349	84.0	74.0	76.0	79.0
TG23	150	229	.315	96.0	83.0	81.0	81.0
TG24	150	229	.306	114.0	99.0	90.0	96.0
TG25	150	229	.268	121.0	106.0	94.0	104.0

(continued)

Table 2. (*continued*)

Girder	d/t	σ_y MPa	σ_{yf} MPa	V P.D. P. KN	V Hoglund KN	V Rockey KN	V Exp KN
G7T1	255	253	259	645.0	575.0	605.0	623.0
TG1′	400	200	280	131.0	119.0	142.0	152.0
TG2′	400	200	280	174.0	136.0	166.0	160.0
TG3′	400	200	280	223.0	168.0	196.0	190.0
TG4′	400	200	280	251.0	191.0	214.0	219.0
TG5′	400	200	280	323.0	256.0	265.0	309.0
PC3	800	216	362	82.0	49.0	73.0	79.0

Table 3. Ratio of predicted to experimental capacity

Investigator	Mean of V_u/V_{ex}	Standard deviation	Range of V_u/V_{ex}
Hoglund	0.91	0.17	0.62–1.17
Rockey	0.97	0.11	0.82–1.22
P.D.P	1.08	0.11	0.86–1.25

6 Conclusion

The finite element method has been used for the large deflection elastic-plastic analysis of plates under shear loading, in the post-buckled regime. This study is focused on the distribution of the stress along the boundaries after buckling in shear, the prediction of the ultimate shear force and the influence of flange strength on the ultimate shear resistance. The following points can be made:

- Prior to buckling, the shear stress is uniformly distributed along the boundary. After buckling, the shear stress distribution becomes non-uniform with a maximum value at the tension corner. Up to the shear force which causes first yielding at the junction between the web and the flange at the tension corner, there is no normal force at the boundaries.
- In most girders used in civil engineering the flanges are light and no useful diagonal tension is developed before the maximum capacity is reached.
- When the d/t ratio of the panel is small enough that the shear force to cause yielding exceeds that to cause elastic bucking, no bending is developed in the flanges. The first yield may take place in the middle of the panel area, due to a combination of membrane stress and bending stresses.
- The proposed model to predict the ultimate load capacity shear webs in a plate girder treats the shear strength of the web and the bending strength of the flanges independently.
- The proposed equation is simple, applies to square panels, give satisfactory results compared with many tests, and avoids the need for iteration used in other methods.

Acknowledgements. The authors wish to express their gratitude to BAU Department of Civil and Environmental Engineering for their support.

References

1. Glassman, J., Moreyra Garlock, M.: A compression model for ultimate postbuckling shear strength. Thin-Walled Struct. **102**, 258–272 (2016)
2. Alinia, M., Gheitasi, A., Shakiba, M.: Postbuckling and ultimate state of stresses in steel plate girders. Thin-Walled Struct. **49**(4), 455–464 (2011)
3. Alinia, M., Shakiba, M., Habashi, H.: Shear failure characteristics of steel plate girders. Thin-Walled Struct. **47**(12), 1498–1506 (2009)
4. Takeda, H., Watanabe, E., Kunishi, R.: Inelastic repetitive shear and flexural buckling of plate girders. Int. J. Struct. Stab. Dyn. **04**(01), 105–124 (2004)
5. Marsh, C.: Theoretical model for collapse of shear webs. J. Mech. Eng. ASCE **108**(5), 819–832 (1982)
6. Basler, K.: Strength of plate girders in shear. ASCE, Proc. No 2967, Part 1, ST 7, pp. 151–180 (1969)
7. Marsh, C.: Photoelastic study of postbuckled shear webs. Can. J. Civ. Eng. **12**(2), 415–417 (1985)
8. Marsh, C., Ajam, W.: Post-buckling strength of shear webs without diagonal tension. In: International Conference on Steel & Aluminium Structure, Cardiff, U.K (1987)
9. Berman, J., Bruneau, M.: Plastic analysis and design of steel plate shear walls. J. Struct. Eng. **129**(11), 1448–1456 (2003)
10. de Béjar, L., Younis, M.: Shear strength of open-web plate girders with inclined tension bars. J. Eng. Mech. **117**(11), 2699–2719 (1991)
11. White, D., Barker, M.: Shear resistance of transversely stiffened steel I-Girders. J. Struct. Eng. **134**(9), 1425–1436 (2008)
12. Yoo, C., Lee, S.: Mechanics of web panel postbuckling behavior in shear. J. Struct. Eng. **132** (10), 1580–1589 (2006)
13. Lee, S., Yoo, C., Yoon, D.: Behavior of intermediate transverse stiffeners attached on web panels. J. Struct. Eng. **128**(3), 337–345 (2002)
14. Lee, S., Yoo, C.: Experimental study on ultimate shear strength of web panels. J. Struct. Eng. **125**(8), 838–846 (1999)
15. Rockey, K.C., Skaloud, M.: The ultimate load behavior of plate girders loaded in shear. J. Struct. Engr. **50**(1), 29–47 (1972)
16. Sadao, K.: Ultimate strength of stiffened plate subjected to shear. In: Proceedings of the IABSE Colloquium, London (1971)
17. Tang, K.H., Evans, H.R.: Transverse Stiffeners for plate girders wens an experimental study. J. Const. Steel Res. 4, 253–280
18. Hoglund, T.: Simply supported long thin plate girders without web stiffeners subjected to distributed transverse load. In: Proceedings of the IABSE Colloquium, London (1971)

Author Index

H. Rodrigues et al. (Eds.): GeoMEast 2019, SUCI, pp. 161–162, 2020.
https://doi.org/10.1007/978-3-030-34216-6

The manufacturer's authorised representative in the EU is Springer Nature Customer Service Centre GmbH, Europaplatz 3, 69115 Heidelberg, Germany. If you have any concerns regarding our products, please contact ProductSafety@springernature.com

Printed and bound by CPI Group (UK) Ltd, Croydon, CR0 4YY
05/10/2026
02228152-0010